The Story of
Gurkha VCs

FireStep Books

www.firesteppublishing.com

The Story of
Gurkha VCs

Dedicated to the hillmen of Nepal,
whose loyalty, cheerfulness and comradeship
have been an inspiration both in peace and war.

FireStep Books

FireStep Books
An imprint of FireStep Publishing

Gemini House
136-140 Old Shoreham Road
Brighton
BN3 7BD

www.firesteppublishing.com

First Edition Published May 1993
Second Edition Published March 2006

First published in this format by FireStep Books,
an imprint of FireStep Publishing, in association with
The Gurkha Museum, 2012

ISBN 978-1-908487-33-9

Cover design FireStep Publishing
Typeset by FireStep Publishing

Printed and bound in Great Britain
by CPI Antony Rowe, Eastbourne and Chippenham

CONTENTS

PREFACE

In 1837 the 'Order of Merit' was introduced in the Honourable East India Company as a gallantry award and it continued to be awarded after the conclusion of the Indian Mutiny when the HEIC's forces were taken over by the Crown. In 1902 the award was re-titled the 'Indian Order of Merit', which title is used throughout this book, regardless of when the award was made. This award was available to Indian and Gurkha officers, non-commissioned officers and soldiers. The Honourable East India Company also introduced in 1837 the 'Order of British India' to reward its Indian and Gurkha officers, sometimes for gallantry but usually for outstanding long and meritorious service.

In the British Army the Meritorious Service Medal had been introduced in 1845 to recognise the gallantry of soldiers, followed by the Distinguished Conduct Medal in 1854, but prior to the Crimean War (1854-1856), the most esteemed award for military prowess was The Most Honourable Order of the Bath (Military Division) which was only awarded to senior officers (Majors and above). Junior officers, non-commissioned officers and soldiers could win promotion in the field, or gain distinction by being mentioned in Generals' despatches, but such mentions went mainly to staff officers serving immediately under a General and only very rarely to those officers and men engaged in front line action.

Queen Victoria and Prince Albert recognised that there was no way of adequately rewarding exceptional individual gallantry of either officers or men in our naval and military services, but it was resolved by a Royal Warrant of 29th January 1856, which brought the Victoria Cross into being. The British public greeted it with great enthusiasm.

The Victoria Cross is a decoration awarded to "those officers and men who have served us in the presence of the enemy and shall then have performed some signal act of valour or devotion to their country". The Warrant further stated that it was to be "highly prized"; it was then and remains so to this day. Although the award came into being in 1856, it was not until 1911, on the occasion of the visit of King George V to India, that Gurkha and Indian officers and soldiers became eligible for the award, (the text of the London Gazette promulgating the decision is given on page 9). Until that year the Indian Order of Merit was the most prized honour that a grateful country could bestow on a brave man.

The first VC awarded to a Gurkha was as a result of the "most conspicuous bravery" of Rifleman KULBIR THAPA on 25th September 1915 in France. In all, the Gurkha Brigade has won a total of 26 Victoria Crosses, each with a unique tale of courage and devotion beyond the call of duty. Of these, thirteen have been awarded to British officers and thirteen to Gurkha officers and men.

In the narratives that follow, some explanation of the titles 'Goorkha', 'Gorkha" and 'Gurkha' might be helpful. In the early days, regiments included 'Goorkha' in their title but in 1891 the spelling was officially changed in the Army to, 'Gurkha'.

In 1901 all ten Gurkha regiments of the Indian Army were designated 'Gurkha Rifles'. After Indian Independence in 1947, the Gurkha regiments which were transferred to the British Army, namely the 2nd, 6th, 7th and 10th retained that title; the 1st, 3rd, 4th, 5th, 8th and 9th, which remained part of the Republic of India Army, together with the re-raised 11th were, in 1949 designated 'Gorkha Rifles'. On 1st July 1994 the 2nd King Edward VII's Own Gurkha Rifles (The Sirmoor Rifles), 6th Queen Elizabeth's Own Gurkha Rifles, 7th Duke of Edinburgh's Own Gurkha Rifles, and 10th Princess Mary's Own Gurkha Rifles, were amalgamated to form The Royal Gurkha Rifles.

When reading the narratives, the descriptions of regiments and battalions can be confusing to the uninitiated. Where, for instance, '1/7th Gurkhas' is mentioned, the second number identifies the regiment and the first number denotes the battalion of that regiment; thus in full, 1/7th Gurkhas is the '1st Battalion 7th Gurkha Rifles' known colloquially as the 'First Seventh Gurkhas'.

It should be noted that in the narratives, the text of the citations from the London Gazette have been produced exactly as written at the time and no attempt has been made to correct what may appear to be obvious errors.

ACKNOWLEDGEMENTS

I would like to express my grateful appreciation to the following for their help in the production of this book:

For the accounts on which the narratives are based:
    The Honorary Secretaries of the Regimental Associations of the Gurkha Brigade and The Brigade of Gurkhas

For reading the narratives and for useful suggestions for the First Edition:
    Mrs C A Mason (former Archivist of The Gurkha Museum Winchester)
    Lieutenant Colonel M H Broadway, late R Signals
    Major M J Fuller, late 2 GR and 4 GR

And for the Second Edition:
    Colonel D R Wood, MBE, late 2 GR
    Lieutenant Colonel M C Barrett OBE, late Q G Signals
    Mr G J Edgerley-Harris, Archivist and Assistant Curator The Gurkha Museum

For permission to reproduce the drawings by the late Lieutenant Colonel C G Borrowman:
    The late Mrs G E M Borrowman

For photograph reproduced on page 89:
    Fayer, London

For help and advice in production and in detail of facts:
    The Curator and staff of The Gurkha Museum
    The Director and staff of the Royal Naval Museum, Portsmouth
    The Curator and staff of the Royal Air Force Museum, Hendon
    Mrs M Fudge, sister of Major F G Blaker VC MC
    D F Harding Esq, MA FRAS
    K Kettle Esq
    Lieutenant Commander I C Macintyre, RN
    Brigadier B S Malik, AVSM, (formerly Military Adviser, The High Commission of India, London)
    Major J M Patrick, late 7 GR
    Mrs M Sharpe
    Mrs A Tottle
    N A Tuckett Esq
    Colonel B R Wadhawan, (formerly Deputy Commandant 58 Gorkha Training Centre, Shillong, India)

Maurice Biggs
Late 4th Prince of Wales's Own Gurkha Rifles
and 10th Princess Mary's Own Gurkha Rifles

FOREWORD [TO THE FIRST EDITION]
BY FIELD MARSHAL SIR JOHN CHAPPLE, GCB, CBE

Colonel 2nd King Edward VII's Own Goorkhas
(The Sirmoor Rifles)

I am delighted to contribute a foreword to this publication which forms part of the series of Gurkha Museum booklets, covering different aspects of the history of Gurkhas in British service since 1815.

This book complements the Museum display of VC winners. It contains the verbatim citations for each recipient as well as additional biographical notes. All this has never been brought together before. It is a tribute to the hard work of Maurice Biggs; and in turn it is a tribute to an exceptional band of warriors.

The Victoria Cross is well known to be the highest decoration that can be awarded for gallantry in the face of the enemy. It is awarded without regard to rank. It is not, however, the earliest award for gallantry; nor was it originally awarded posthumously or to soldiers of the Indian Army. For all these reasons the list of recipients contains some anomalies. Before 1911 it contains only the British officers who were serving in command of Gurkhas. Before 1900 it had not been awarded posthumously; thus we do not know how long the list would have been had these two conditions not applied!

Further to this, for the first 50 years or so since its institution, there were very few alternative awards available either for British or Gurkha officers or soldiers. This can be illustrated by the table attached, which shows how many different types of honours and awards (including those for merit rather than bravery) were available to Gurkha officers and soldiers since 1815. This list lengthens progressively but it illustrates the point that up to 1907 there was only the Indian Order of Merit (IOM) available as a gallantry decoration for soldiers of the Indian Army.

A word of explanation is needed about the Regiment in which the British officers served. Between the late 1850s and the turn of the century, officers were gazetted to the Staff Corps of one of the Presidency Armies (or later to the Indian Staff Corps). This was a technicality since they served (and were shown in the Army List) in the regiment to which they were 'attached'. After about 1900, they were gazetted directly to their parent regiment. Conversely, two officers who won the VC in World War 2 (Captain M Allmand and Major F G Blaker MC), were both commissioned into other regiments and were 'attached' in the sense of being seconded to the Gurkha regiment in which they were serving at the time they won their award.

London
May 2003

FOREWORD TO THE SECOND EDITION
BY FIELD MARSHAL SIR JOHN CHAPPLE, GCB, CBE, DL

Vice Patron of The Gurkha Museum

This book has proved to be popular enough to run to two reprints and when the last of these ran out we thought it a suitable opportunity to revise and update it. Further discoveries during the course of research, and the deaths of some of the World War 2 recipients of the Victoria Cross have made this step desirable. Moreover it was thought to be fair and useful to add the details of three British Officers and two Gurkha Riflemen who were awarded the Albert Medal, an award that was later to be exchanged for the George Cross by those who survived.

The additional and editorial work for this enlarged second edition has been done mostly by Lieutenant Colonel Mike Barrett, OBE, a Trustee of The Gurkha Museum to whom we are greatly indebted. A Member of The Orders and Medals Research Society, his expertise has been applied to the maintenance and display of the Museum's comprehensive collection of Orders, Decorations and Medals.

The collection includes the only George Cross, the only Empire Gallantry Medal, one of the two Albert Medals awarded to Gurkhas and ten of the Victoria Crosses featured in this book.

John Chapple

London
January 2006

13

GALLANTRY AND MERIT AWARDS FOR GURKHAS

Period	Gurkha Officers Eligible for	Gurkha Other Ranks Eligible for
1815 - 1837	None (monetary awards and promotion in the field only)	
1837 - 1843	OBI 1st & 2nd Class, IOM 1st, 2nd & 3rd Class	IOM 1st, 2nd & 3rd Class
1843 - 1888	OBI 1st & 2nd Class, IOM 1st, 2nd & 3rd Class AM (from 1877), Mention	IOM 1st, 2nd & 3rd Class, AM (from 1877) IMSM (from 1888), Mention
1888 - 1907	OBI 1st & 2nd Class, IOM 1st, 2nd & 3rd Class, AM, Mention, RVM (from 1896)	IOM 1st, 2nd & 3rd Class AM, IMSM, Mention, RVM (from 1896)
1907 - 1911	OBI 1st & 2nd Class, IOM 1st, 2nd & 3rd Class, IDSM (from 1907), AM, Mention, RVM	IOM 1st, 2nd & 3rd Class, IDSM (from 1907), AM, Mention, IMSM, RVM
1911 - 1914	VC, OBI 1st & 2nd Class, IOM (new) 1st & 2nd Class, AM, IDSM, Mention, RVM	VC, IOM (new) 1st & 2nd Class, AM, IDSM, IMSM, Mention, RVM
1914 - 1920	VC, OBI 1st & 2nd Class, IOM 1st & 2nd Class, MC (from 1914), AM, IDSM, MID, RVM	VC, IOM 1st & 2nd Class, AM, IDSM, IMSM, MID, RVM
1920 - 1940	VC, OBI 1st & 2nd Class, IOM 1st & 2nd Class, MC, AM, IDSM, MID, RVM	VC, IOM 1st & 2nd Class, AM, IDSM, EGM (from 1922), MID, IMSM, RVM
1940 - 1944	VC, GC, MBE, OBI 1st & 2nd Class, OB, IOM 1st & 2nd Class, MC, AM, GM, IDSM, BGM, MID & Commendation, RVM	VC, GC, IOM 1st & 2nd Class, AM, GM, IDSM, BGM, MID & Commendation, RVM, BEM, IMSM
1944 - 1947	VC, GC, MBE, OBI 1st & 2nd Class, OB, IOM (new) single Class, MC, AM, GM, BGM, MID & Commendations, RVM	VC, GC, IOM (new) single Class, AM, GM, IDSM, BGM, MM, MID & Commendations, RVM, BEM, IMSM
1948 - 1974	VC, GC, MVO, MBE, MC, AM (until 1971), GM, MID & Commendations	VC, GC, AM (until 1971), DCM, GM, MM, MID & Commendations, RVM, BEM

Period	Gurkha Officers Eligible for	Gurkha Other Ranks Eligible for
1974 - 1993	VC, GC, MVO, MBE, MC, GM, QGM, MID & Commendations	VC, GC, DCM, GM, MM, QGM, MID & Commendations, RVM, BEM
1993 - 2001	VC, GC, MVO, MBE, CGC, MC, GM, QGM, MID, Queen's Commendations, Commendations	VC, GC, MBE, CGC, MC, GM, QGM, Mention, Queen's Commendations, Commendations, RVM
2002 onwards	VC, GC, MVO, MBE, CGC, MC, GM, QGM, MID, Queen's Commendations, Commendations	VC, GC, MBE, CGC, MC, GM, QGM, MID, Queen's Commendations, Commendations, RVM, MSM (from 2002)

ABBREVIATIONS USED IN THE TABLE

AM	Albert Medal
BEM	British Empire Medal
BGM	Burma Gallantry Medal
CGC	Conspicuous Gallantry Cross
Commendation	Commendations awarded by a Senior Authority e.g. Commander-in-Chief, General Officer Commanding, Force commander
DCM	Distinguished Conduct Medal
EGM	Empire Gallantry Medal
GC	George Cross
GM	George Medal
IDSM	Indian Distinguished Medal
IMSM	Indian Meritorious Service Medal
IOM	Indian Order of Merit
MBE	Member *(5th Class)* of The Most Excellent Order of the British Empire
MC	Military Cross
Mention	Mention *(in Despatches)*
MID	Mention in Despatches
MM	Military Medal
MSM	Meritorious Service Medal
MVO	Member *(5th Class)* of The Royal Victorian Order
OB	Order of Burma
OBI	Order of British India
QGM	Queen's Gallantry Medal
Queen's Commendation	Awards made by the Sovereign; QCBC Queen's Commendation for Brave Conduct QCVS Queen's Commendation for Valuable Service
RVM	Royal Victorian Medal
VC	Victoria Cross

GLOSSARY

ADC	Aide-de-Camp
FRCS	Fellow of the Royal College of Surgeons
GCO	Gurkha Commissioned Officer. A post 1948 rank in The [British] Brigade of Gurkhas
GHQ	General Headquarters
GR	Gurkha Rifles
GSO 1	General Staff Officer 1st Grade
Havildar	Indian Army rank equivalent to Sergeant
Havildar Major	Indian Army rank equivalent to Sergeant Major
HMG	His (or Her) Majesty's Government
ICS	Indian Civil Service
IMS	Indian Medical Service
Jemedar (or Jemadar)	Indian Army rank equivalent to Lieutenant (QGO)
Lance Naik	Indian Army rank equivalent to Lance Corporal
LMG	Light Machine Gun
MMG	Medium Machine Gun
Naik	Indian Army rank equivalent to Corporal
Nullah (or nala)	Dried river bed
PIAT	Projector Infantry Anti-Tank
Piquet (or picquet)	A small detachment of troops positioned to give early warning of attack
QGO	Queen's Gurkha Officer. A post 1952 rank in The [British] Brigade of Gurkhas, equivalent to the Indian Army Viceroy's Commissioned Officer
RE	Royal Engineers
RFC	Royal Flying Corps
Sangar (or sungar)	Stone breastwork in a (*usually*) defensive position
Sepoy	A private soldier in the Indian Army
Subedar (or Subadar)	Indian Army rank equivalent to Captain (QGO)
Subedar Major	Indian Army rank equivalent to Major (QGO), the senior Gurkha Officer in a Gurkha battalion
TSMG (or TMG)	Thompson sub-machine gun

ORDERS AND DECORATIONS

VC	Victoria Cross
GC	George Cross
GCB	Knight Grand Cross, The Most Honourable Order of the Bath
KCB	Knight Commander, The Most Honourable Order of the Bath
CB	Companion, The Most Honourable Order of the Bath
GCMG	Knight Grand Cross, The Most Distinguished Order of St. Michael and St. George
CMG	Companion, The Most Distinguished Order of St. Michael and St. George
CIE	Companion, of The Most Eminent Order of the British Empire
CVO	Commander, The Royal Victorian Order
CBE	Commander, The Most Excellent Order of the British Empire
DSO	Distinguished Service Order
MVO	Member, The Royal Victorian Order
OBI	Order of British India
IOM	Indian Order of Merit – Military
MC	Military Cross
AM	Albert Medal
KPM	King's Police Medal
MM	Military Medal
EGM	Empire Gallantry Medal

Extract from
The London Gazette
Published by Authority
TUESDAY, 12 DECEMBER, 1911

War Office,
12ᵗʰ December, 1911.

ROYAL WARRANT

GEORGE R.I.
Warrant extending the Decoration of the Victoria Cross to the Native Officers, Non-commissioned Officers and Men of the Indian Army.

GEORGE, by the Grace of God, of the United Kingdom of Great Britain and Ireland, and of the British Dominions beyond the Seas, King, Defender of the Faith, Emperor of India, to all to whom these presents shall come, Greeting.

WHEREAS Her Majesty, Queen Victoria, by a Warrant under Her Royal Sign Manual, countersigned by one of Her Principal Secretaries of State, and bearing date at Her Court at Buckingham Palace, the twenty-ninth day of January, one thousand eight hundred and fifty-six, in the nineteenth year of Her reign, did institute and create a new naval and military decoration, to be styled and designated the "Victoria Cross," which decoration She expressed Her desire should be highly prized and eagerly sought after by the Officers and men of Her Naval and Military Services, and did also make, ordain, and establish the rules and ordinances therein set forth for the government of the same, to be thenceforward inviolably observed and kept.

And whereas for divers reasons Us thereunto moving, We are desirous of rewarding the individual gallant services of native officers, non-commissioned officers and men of Our Indian Army by the bestowal of the said decoration, which We are desirous shall be highly prized and eagerly sought after by the said native officers, non-commissioned and men.

Now know ye that We, of Our especial grace, certain knowledge and mere motion, have thought fit hereby to signify Our Royal Will and Pleasure that the said decoration shall be conferred on the native officers, non-commissioned officers and men of Our Indian Army who may be qualified to receive the same, in accordance with the rules and ordinances made, ordained and established for government thereof by the said recited Warrant, and We do by these Presents, for Us, Our Heirs and Successors, ordain and appoint that it shall be competent for the native officers, non-commissioned officers and men of Our Indian Army to obtain the said decoration in the manner set forth in the rules and ordinances referred to, or in accordance with any further rules and ordinances which may hereafter be made and promulgated by Us, Our Heirs and Successors, for the government of the said decoration.

And We do further, for Us, Our Heirs and Successors, ordain and appoint that in place of the special pension conferred by the fourteenth rule of the said recited Warrant, every native officer who shall have received the Cross shall from the date of the act by which such decoration has been gained be entitled to a special pension of five hundred and twenty-five rupees a year, and each additional bar conferred under the fourth rule on such native officer shall carry with it an additional pension of one hundred and fifty rupees a year. In the case of warrant or non-commissioned officer or soldier the special pension shall be one hundred and fifty rupees, with seventy-five rupees additional for each additional bar. On the death of a recipient of the Cross these pensions shall continue to his widow until her death or re-marriage.

Given at Our Court of St. James's this 21ˢᵗ day of October, in the second year of Our Reign, and in the year of Our Lord one thousand nine hundred and eleven.

By His Majesty's Command,
HALDANE OF CLOAN.

Victoria Cross Awards

The Story of Gurkha VCs

LIEUTENANT J A TYTLER VC

Bengal Staff Corps attached 66th or Goorkha Regiment
later 1st King George V's Own Gurkha Rifles (The Malaun Regiment)

Date of Action	*Campaign*
10th February 1858	*Indian Mutiny*

John Adam Tytler was born at Monghyr, Bengal, on 29th October 1825, the third son of Dr John Tytler, a surgeon in the service of the East India Company, and his wife Anne, daughter of W Gillies Esq, of London. When only 5 years old, John Tytler was sent home to England into the care of his mother's sisters, with whom he lived until his parents returned from India in 1835.

The family stayed for a year in London and then went to Jersey, where John attended a day school near St. Helier. In 1837, his father died and he and his mother went to live in Edinburgh, where John attended College.

On the recommendation of his father's old friend, General Sir Jeremiah Bryant, a Director of the East India Company, Tytler was Commisioned on 10th December 1844 in the 1st Bengal Native Infantry. On 14th March 1845 he was posted to the 66th Native Infantry and first saw active service on the Peshawar Frontier in 1851 under Sir Colin Campbell. On 27th February 1850, his Regiment became the 66th or Goorkha Regiment of Native Infantry – later to become the 1st King George V's Own Gurkha Rifles (The Malaun Regiment).

Tytler was appointed Adjutant in October 1853 and, during and immediately after the Mutiny of 1857, saw service in the hills around Naini Tal. On 17th September 1857, with seventy men of the 66th, he played a major part in the defeat of some one thousand rebel horse and foot attacking Haldwani. In early February 1858, the 66th were again deployed at Haldwani against two large rebel forces. On 9th February 1858 Lieutenant Tytler was with five hundred men of the 66th, which, with two hundred and ten other infantry, two hundred cavalry, and two six-pounder guns, surprised a rebel force of between four and five thousand infantry, more than one thousand cavalry and four guns, before Charpura. The next day, two companies of the 66th, under the command of Captain Ross, advanced steadily against the rebel right flank in the face of heavy fire from the enemy guns and it was in this attack that Lieutenant Tytler distinguished himself.

The citation in the London Gazette of 24th August 1858 read:

" On the attacking parties approaching the enemy's position under a heavy fire of round shot, grape, and musketry, on the occasion of the Action at Choorpoorah, on the 10th February last , Lieutenant Tytler dashed on horseback ahead of all, and alone, up to the enemy's guns, where he remained engaged in hand to hand, until they were carried by us; and where he was shot through the left arm, had a spear wound in his chest, and a ball through the right sleeve of his coat. [Letter from Captain C. C. G. Ross, commanding 66th (Goorkha) Regiment, to Captain Brownlow, Major of Brigade, Kemaon Field Force]".

For this act of valour, Lieutenant Tytler was awarded the Victoria Cross, the first to be won by an officer of a Goorkha Regiment.

In July 1862 Tytler was appointed Second in Command of the 3rd Goorkha (The Kemaoon) Regiment and in November 1863 joined the 4th Goorkha Regiment as acting Commandant during the Ambeyla campaign becoming permanent Commandant of the Regiment in December 1865. The 4th Goorkha Regiment had been raised in 1857 from a draft of officers and men from the 1st Goorkha Regiment and he remained in command of it for seventeen years. He took part in the Hazara [Black Mountain] Expedition of 1868, was twice Mentioned in Despatches, and for services in the Looshai Campaign, was appointed a Companion of the Order of The Bath in 1872. In 1878 he commanded a Brigade on the North West Frontier with the rank of Brigadier General.

He died of pneumonia at Thal in the Kurram Valley on 1st February 1880, aged 54.

MAJOR D MACINTYRE VC

Bengal Staff Corps attached 2nd Goorkha (The Sirmoor Rifle) Regiment
later 2nd King Edward VIIs Own Gurkha Rifles (The Sirmoor Rifles)

Date of Action	*Campaign*
4th January 1872	*Looshai*

Donald Macintyre was born on 12th September 1831 at Kincraig House, Ross-shire, Scotland and married Angelica, the daughter of the Reverend T J Patteson of Kinettles, Forfar. They had three sons, Donald, Frank and Ian. Donald served in 2nd King Edward's Own Gurkha Rifles (The Sirmoor Rifles) from August 1905 to May 1919, while Frank served in the Royal Naval Air Service in the 1914-1918 war. Ian became a Captain in the Royal Navy and was awarded the CB, CBE and DSO in the 1939-1945 war, serving as Chief of Staff to Admiral Sir Max Horton during the Battle of the Atlantic and commanding the anti-aircraft cruiser HMS Scylla on several Russian convoys. He ended the war as Captain of HMS Indefatigable, an aircraft carrier in the Pacific.

Donald Macintyre (Senior) was educated at private schools in England and abroad before attending the Addiscombe Military Seminary. He was commissioned from there in June 1850 and on arrival in India was posted to the 66th Regiment of Native Infantry (Goorkhas) with which Regiment he served until 1857.

During this time he took part in various frontier campaigns, including the Kurram Expedition to Afghanistan in 1856 under Sir Neville Chamberlain. On 6th August 1857, he was appointed to raise "The Extra Goorkha Regiment" (later 4th Prince of Wales's Own Gurkha Rifles) and on 2nd November 1858 was posted to the Sirmoor Rifle Regiment, becoming Adjutant in May 1860. In February 1861, he was appointed to the Bengal Staff Corps and became Second-in-Command of the battalion in February 1864. After various campaigns against Mohmands and other hill tribes, he accompanied the battalion to the Hazara country in May 1868 and in 1872, as a Major, served with the battalion in the first Looshai Expedition, where he was awarded the Victoria Cross in the storming of the stockaded village of Lalgnoora. He was also Mentioned in Despatches, promoted to the brevet rank of Lieutenant Colonel and given the thanks of the Governor General of India.

The citation in the London Gazette of 27th September 1872 read as follows:-

"For his gallant conduct at the storming of the stockaded village of Lalgnoora on the 4th January, 1872.

Colonel Macpherson C.B., V.C.*, Commanding the 2nd Goorkha Regiment, in which Lieutenant-Colonel Macintyre was serving at the time as second-in-command, reports that this Officer, who led the assault, was the first to reach the stockade (on this side from 8 to 9 feet high): and that to climb over and disappear among the flames and smoke of the burning village, was the work of a very short time. The stockade, he adds, was successfully stormed by this Officer under fire, the heaviest the Looshais delivered that day".

Rifleman Inderjit Thapa, who immediately followed Major Macintyre over the stockade received the Indian Order of Merit, 3rd Class. Lieutenant Colonel Macintyre was appointed Commanding Officer in April 1876 and promoted Colonel in September of that year. He was present with the battalion at Delhi during HRH The Prince of Wales's visit on 11th January 1876 and commanded the battalion during its tour of duty in Malta and Cyprus from 1st June 1878 to 8th October 1878. On the return of the battalion to India he again took it on service with the Peshawar Valley Field Force. He retired on 24th December 1880, being appointed Honorary Major General on that date.

General Macintyre was a great shikari and was author of two books, "Hindu Koh" and "Wanderings and Wild Sport on, and beyond the Himalayas" as well as contributing to the "Encyclopaedia of Sport" published in 1898. He was a Justice of the Peace for Ross-shire and a Fellow of the Royal Geographical Society.

In retirement he lived at Mackenzie Lodge, Fortrose, Ross-shire and belonged to the United Services Club, London, and the Highland Club, Inverness. His recreations were wild sport, golf and cycling. He died on 15th April 1903 at Fortrose aged 71 and is buried there, in the family plot at Rosemarkie churchyard.

*N.B. In 1872, the CB took precedence over the VC.

CAPTAIN G N CHANNER VC

Bengal Staff Corps attached 1st Goorkha Regiment
later 1st King George V s Own Gurkha Rifles
(The Malaun Regiment)

Date of Action   *Campaign*
20th December 1875   *Perak*

George Nicholas Channer was born at Allahabad, India, on 7th January 1843, the son of Colonel George Girdwood Channer, and his wife Susan, daughter of the Reverend Nicolas Kendall, Vicar of Lanlivery, Cornwall. He was educated at Truro Grammar School and Cheltenham College.

George Channer was commissioned on 4th September 1859 and became an 'Ensign Unattached' until being attached to the 89th Foot from 1862 – 1864. He had been promoted to Lieutenant on 28th August 1861 and served on the North West Frontier in 1863 – 64. In March 1867 he was appointed Quartermaster of the 35th (Manipuri) Regiment of Bengal Native Infantry and was transferred to the 2nd Bengal Native (Light) Infantry in the same capacity in December 1869. Promoted Captain on 4th September 1871, he served in the Looshai campaign before being appointed to the 1st Goorkha Regiment on 6th March 1873.

For a conspicuous act of bravery on 20th December 1875, at the Bukit Pass and stockades in Negri Sembilan, Malay Peninsula, Captain Channer was awarded the Victoria Cross and promoted to Brevet Major.

The citation in the London Gazette of 14th April 1876 stated:

"For having, with the greatest gallantry, been the first to jump into the Enemy's Stockade, to which he had been dispatched with a small party of the 1st Ghoorkha Light Infantry, on the afternoon of the 20th December, 1875, by the Officer commanding the Malacca Column, to procure intelligence as to its strength, position, etc.,

Major Channer got completely in rear of the Enemy's position, and finding himself so close that he could hear the voices of the men inside, who were cooking at the time, and keeping no look out, he beckoned to his men, and the whole party stole quietly forward to within a few paces of the Stockade. On jumping in, he shot the first man dead with his revolver, and his party then came up, and entered the Stockade, which was of a most formidable nature, surrounded by a bamboo palisade; about seven yards within was a loghouse, loop-holed, with two narrow entrances, and trees laid latitudinally, to the thickness of two feet.

The Officer commanding reports that if Major Channer, by his foresight, coolness, and intrepidity, had not taken this Stockade, a great loss of life must have occurred, as from the fact of his being unable to bring guns to bear on it, from the steepness of the hill, and the density of the jungle, it must have been taken at the point of the bayonet".

Major Channer took part in the Jowaki Afridi Expedition of 1877/78 and was Mentioned in Despatches and promoted to Brevet Lieutenant Colonel during the Afghan War of 1878 – 80. He was appointed a Companion of the Order of The Bath and again Mentioned in Despatches whilst commanding the 1st Brigade on the Black Mountain Expedition of 1888. He was promoted Major General on 27th April 1893 and Lieutenant General on 9th November 1896. He died at Westward Ho, Devon on 13th December 1905, aged 62 and is buried in the East the Water Cemetery at Bideford.

CAPTAIN J COOK VC

Bengal Staff Corps attached 5th Goorkha Regiment
(The Hazara Goorkha Battalion)
later 5th Royal Gurkha Rifles (Frontier Force)

Date of Action	*Campaign*
2nd December 1878	*Afghanistan 1878-80*

John Cook was born in Edinburgh on 28th August 1843. He was commissioned on 19th December 1860 and became an 'Ensign Unattached' before joining the 4th Bengal Europeans and then Her Majesty's 107th Foot in 1861/62. He was appointed to the 3rd Sikh Infantry of the Punjab Frontier Force and took part in the Ambeyla Campaign of 1863 in which he was Mentioned in Despatches. He was made Adjutant in November 1864 and in 1868 took part in the Hazara [Black Mountain] Expedition. Promoted Captain in 1872 he joined the 5th Goorkha Regiment (The Hazara Goorkha Battalion) on 27th March 1873 and was appointed a Wing Officer.

The onset of hostilities of the 2nd Afghan War in November 1878 saw the 5th Goorkhas forming part of the Kurram Field Force, one of three Columns under the command of Major General Frederick Roberts (later to become Field Marshal Earl Roberts of Kandahar VC), preparing to invade Afghanistan via the Kurram Valley and Thal.

On reaching the Afghan fort at Kurram, which was found abandoned, reconnaisance showed that the Afghan Army (numbering some eighteen thousand men with eleven guns) was preparing defensive positions on the Peiwar Kotal. This mountain feature rose some nine thousand four hundred feet above sea level and averaged two thousand feet above the floor of the valley approach and the Spingawai Pass. It lay to the east of the main enemy position.

To turn the Afghan defences, General Roberts ordered part of his Force to move under cover of darkness via the Spingawai approach. With the 5th Goorkhas in the van, and just before first light, the enemy look-outs became alerted and opened fire. The advance guard of 5th Goorkhas immediately formed up from column of fours into a company line and led by Major Fitzhugh and Captain Cook, rushed straight at a barricade which, as dawn developed, they saw some fifty yards to their front. The remainder of the Regiment now joined by the 72nd Highlanders (later Seaforth Highlanders), swarmed round the flanks of the obstacle and together charged up the steep and thickly wooded hillside, driving the enemy headlong from their formidable defence works. Daylight had hardly broken when Captain Cook found himself confronted by a considerable number of the enemy who were attempting to save one of their guns. He collected a few men who charged the Afghans who fled, leaving many dead.

During this operation, Captain Cook saved the life of Major Galbraith, Assistant Adjutant-General, by engaging and throwing down an Afghan about to shoot him.

For these acts of bravery and for his gallant leadership in the assault, he was awarded the Victoria Cross. The citation, contained in the London Gazette of 18th March 1879, read as follows:

"For a signal act of valour at the action of the Peiwar Kotal on the 2nd December, 1878, in having, during a very heavy fire, charged out of the entrenchments with such impetuosity that the enemy broke and fled, when, perceiving, at the close of the melee, the danger of Major Galbraith, Assistant Adjutant-General, Kurum Column Field Force, who was in personal conflict with an Afghan soldier, Captain Cook distracted his attention to himself, and aiming a sword cut which the Douranee avoided, sprang upon him, and, grasping his throat, grappled with him.

They both fell to the ground. The Douranee, a most powerful man, still endeavouring to use his rifle, seized Captain Cook's arm in his teeth, until the struggle was ended by the man being shot through the head".

He received his Victoria Cross from General Roberts during a parade held in honour of the Queen's Birthday at Ali Khel on 24th May 1879.

Captain Cook displayed further acts of gallantry during the action at Monghyr Pass in December 1878 and at Charasia when, after the massacre at the British Residency, Kabul, on 3rd September 1879, the Force, now renamed the Kabul Field Force, commenced its advance upon Kabul. At Charasia some eleven miles from Kabul, strong opposition was encountered in precipitous and difficult terrain. Again Captain Cook, in command of two companies of the 5th Goorkhas, alongside their old comrades of the 72nd Highlanders, overwhelmed the enemy position. He was promoted to Major on 22nd November 1879 for his services in this campaign.

After a brief respite in the shelter of the defences at Sherpur, just north of Kabul, Cook's companies were despatched as part of a Force to re-capture the Takht-I-Shah massif, which dominated the southern approaches to Kabul and was held by the Afghans in great strength. Leading his men towards their objective on 13th December 1879, Major Cook fell seriously wounded, died of his wounds a week later and was buried in the Sherpur Cantonment cemetery.

In his despatch of 23rd February 1880, Lieutenant General Sir Frederick Roberts wrote:

"By Major Cook's death. Her Majesty has lost the services of a most distinguished and gallant officer and the Kabul Field Force a comrade who one and all honoured and admired".

Many years later, Parsu Khattri, then the Subedar Major of the 5th Goorkhas and himself described as "the bravest of the brave", said of John Cook, "He was the bravest man I have ever seen, braver even than Roberts Sahib Bahadur, whom all the Regiment considered very brave, above all other men".

LIEUTENANT R K RIDGEWAY VC

Bengal Staff Corps attached 44th (Sylhet) Light Infantry
later 1st Battalion 8th Gurkha Rifles

Date of Action	*Campaign*
22nd November 1879	*Naga 1879-80*

Richard Kirby Ridgeway was born at Oldcastle, County Meath, Ireland, on 18th August 1848. The son of a surgeon, he was educated at a private school and later at the Royal Military College, Sandhurst. He was gazetted into the 96th Regiment in January 1868, but in 1872 he transferred to the 43rd (Assam) Regiment of Bengal Native (Light) Infantry. On 3rd July 1874 he became Adjutant of the 44th (Sylhet) Light Infantry, an appointment he held until 1880.

For much of its existence, the 44th had been involved in operations against unruly tribes in the north-east of India, one of these being the first Naga Hills expedition in 1875 and for his conduct during this campaign. Lieutenant Ridgeway was Mentioned in Despatches. In 1879, the Naga tribesmen were again causing trouble and it was decided to launch a further expedition into the Naga Hills using the 44th and the 43rd Regiments, [the latter later became the 2nd Battalion, 8th Gurkha Rifles].

The stronghold of the Nagas was the village of Konoma. This had been strongly fortified in a series of terraces, each with stone walls and towers. On 22nd November, the 44th made an attack on the Konoma position and succeeded in capturing the outlying fortifications. However, the inner lines, which consisted of a stone-faced escarpment, surmounted by a loop holed stockade, in all about twelve feet high and defended by several thousand well-armed Nagas, proved to be intractable. In an attempt to break down the stockade, the Regiment's two seven-pounder mountain guns were brought within twenty yards, but this was only partially successful. Two assaults were then made on the stockade, both led by Lieutenant Ridgeway. On the second assault, under very heavy enemy fire, he ran up to the stockade and, although severely wounded in the left shoulder, he tore down the planking to create an entrance into the stockade, and heroically remained there until his men were able to get through it.

Lieutenant Ridgeway was promoted Captain on 8th January 1880. For his gallantry at Konoma, he was awarded the Victoria Cross, promulgated in the London Gazette of 11th May 1880; the citation read:

"For conspicuous gallantry throughout the attack on Konoma, on the 22nd November, 1879, more especially in the final assault, when, under a heavy fire from the enemy, he rushed up to a barricade and attempted lo tear down the planking surrounding it, to enable him to effect an entrance, in which act he received a very severe rifle shot wound in the left shoulder".

In 1883, Captain Ridgeway attended the Staff College and after two staff appointments in India, he was given command of his Regiment, then re-titled the 44th Gurkha (Rifle) Regiment of Bengal Infantry, in 1891. In that same year, the Regiment took part in the Manipur Expedition and Major Ridgeway was again Mentioned in Despatches. He was promoted to Lieutenant Colonel in 1894 and took part in the Tirah Expedition in 1897, before being promoted Colonel in 1898. He was placed on the Unemployed Supernumerary List in the same year and was appointed a Companion of the Order of The Bath in 1905.

He retired in 1906 and returned with his wife Emily to live in Harrogate in Yorkshire, where he died on 11th October 1924 at the age of 76. Cremated in Leeds, his ashes were scattered in the Cemetery Copse near Columbarium on 25th October.

LIEUTENANT C J W GRANT VC

Indian Staff Corps attached 12th Regiment (2nd Burma Battalion),
Madras Infantry, commanding a detachment of the 43rd Gurkha (Rifle)
Regiment of Bengal Infantry
later 2nd Battalion 8th Gurkha Rifles

Date of Action	*Campaign*
27th March 1891	*North East Frontier*

Charles James William Grant was born at Bourlie, Aberdeenshire, Scotland on 14th October 1861 the elder son of Lieutenant General D G S St J Grant and Helen, daughter of Colonel William Bisset. He was educated privately and then entered the Royal Military College, Sandhurst prior to being commissioned into The Suffolk Regiment on 10th May 1882. Two years later he transferred to the Indian Staff Corps and between 1885 and 1887 served in the Burma Expedition. At the beginning of 1891 he was in command of a small garrison in Tamu, on the Assam/Burma border.

At the end of March 1891, the little independent state of Manipur on the North-east Frontier of India rose in rebellion, removed the ruler acknowledged by the Government of India, and set up another in his place. Mr Quinton, the British Commissioner in Assam, went there with four civil officers and an escort of seven British officers and 454 Gurkhas from Kohima to arrest the rebel leaders and restore the situation.

The Manipuris however had no intention of giving in and put up stiff resistance for two days. They then sounded the "Cease Fire" and suggested a truce. This was agreed to and it was arranged that the Commissioner and other officers should meet the Manipuris' leaders unarmed to discuss terms. Treachery and murder followed. The Commissioner, the Political Officer, the Colonel commanding the Gurkha escort, and three other officers were seized and barbarously put to death. The escort, resting in the Residency grounds and unaware of what had happened, was then attacked by overwhelming numbers in the darkness, and having expended most of its ammunition, had to withdraw as best it could, some men towards Silchar and others towards Kohima.

The Resident's permanent escort consisted of one company of 43rd Gurkha Rifles of which one platoon (one Gurkha officer and thirty other ranks) was detached at Lang Thobal, about four miles from the Residency. On hearing from the withdrawing soldiers what had happened and that the rebels were advancing on his post, the Gurkha officer, Jemedar Birbal Nagarkoti, decided to withdraw to the nearest military post, commanded by Lieutenant Grant at Tamu, about forty miles away across very rough and hilly country.

Having heard the news from Jemedar Birbal Nagarkoti, Lieutenant Grant immediately telegraphed for permission to go to the assistance of the beleaguered garrison at Manipur and, having received it, proceeded on the morning of 28th March towards Manipur, some fifty miles away, with fifty men of the Madras Infantry (2nd Burma Battalion), the platoon of the 43rd Gurkha Rifles, and three pack elephants.

Progress was slow, about one mile an hour, but they reached Kongaung early on 29th March and drove two hundred men of the Manipur army out of Palel early on the 30th. They reached Lang Thobal, about fifteen miles from Manipur, later that day and found it had been occupied by a large force of Manipuris, well entrenched behind a shallow river. Lieutenant Grant's force crept to within a hundred yards of the enemy and then made a bayonet charge across the river. During this attack Lieutenant Grant was wounded. The Manipuris gave way but rallied at a second line a few hundred yards back. They finally ran in full retreat before Lieutenant Grant's force of eighty men to a range of hills a mile away. It was later learned that there were over eight hundred Manipuris manning the position at Lang Thobal.

Lieutenant Grant now set about improving the defences at Lang Thobal. He realised that there was little hope of his force reaching Manipur, owing mainly to shortage of ammunition, so he determined to hold the position until relieved. The first counter-attack came the following morning but the accurate fire from the Gurkha soldiers beat this off. Further attacks followed, supported by two field guns, but these too were beaten off. The next day the Manipur leaders sent a message together with a bribe of rations in an attempt to get Lieutenant Grant to withdraw, but he returned them saying he would only do so with a member of the Manipuri royal family as hostage. Incensed, the Manipuris built up their force to two thousand strong, put in a continuous bombardment and counter-attacked on all sides. But they failed to capture Lang Thobal. Finally on 9th April 1891, Lieutenant Grant received orders to withdraw to Tamu, which he succeeded in doing without incurring further casualties.

For his action at Lang Thobal, Lieutenant Grant was promoted to Captain on 10th May, to Brevet Major on 11th May, and on 26th May was awarded the Victoria Cross. Jemedar Birbal Nagarkoti was awarded the Order of British India and every survivor of his party the Indian Order of Merit.

The citation in the London Gazette of 26th May 1891 read as follows:

"For the conspicuous bravery and devotion to his country displayed by him in having, upon hearing on the 27th March, 1891, of the disaster at Manipur, at once volunteered to attempt the relief of the British Captives, with 80 Native Soldiers, and having advanced with the greatest intrepidity, captured Thobal, near Manipur, and held it against a large force of the enemy.

Lieutenant Grant inspired his men with equal heroism, by an ever-present example of personal daring and resource".

In February 1906, Grant, now a Lieutenant Colonel, was appointed to command the 89th Punjabis, and in 1907, the 92nd Punjabis. He retired as a Brevet Colonel in 1910 but, as a Retired Officer during World War 1 he served as a Draft Conducting Officer, graded as Staff Lieutenant 2nd Class on 8th May 1915. After the war he returned to England where he died in Sidmouth, Devon, on 23rd November 1932 at the age of 71 and was buried there, in the Temple Road cemetery.

LIEUTENANT G H BOISRAGON VC

Indian Staff Corps attached 5th Gurkha (Rifle) Regiment
later 5th Royal Gurkha Rifles (Frontier Force)

Date of Action	*Campaign*
2nd December 1891	*Hunza 1891*

Guy Hudleston Boisragon was the representative of a very ancient Huguenot family. His ancestor, Louis Chevelleau de Boisragon, escaped to Holland after the Edict of Nantes, and took service with Schornberg's Horse. Subsequently he came to England with William of Orange, fought at the Battle of the Boyne and married Marie Henriette, daughter of the Marquis de Rambouillet, who was herself a refugee to England.

Guy Boisragon was born at Kohat, Punjab on 5th November 1864, the eldest son of Major General H F M Boisragon and Anna, daughter of William Hudleston ICS, Madras. He was educated at Charterhouse, and in 1885 was gazetted from Sandhurst to The Lincolnshire Regiment. His father had already had a distinguished military career and was the founder of the 5th Royal Gurkha Rifles (Frontier Force) which was raised in 1858 as the 25th Punjab Infantry or The Huzara Goorkha Battalion.

Guy Boisragon followed in his father's footsteps and on 7th April 1887 was appointed as a Lieutenant to the 1st Battalion, 5th Goorkha Regiment, The Hazara Goorkha Battalion. His first battle experience came during the second Black Mountain Expedition from September to November 1888, when his battalion was in action against the tribesmen of the North West Frontier. In 1891, he again saw action in the second Miranzai Expedition and the capture of Samana.

Towards the end of 1891, the Hunzas and Nagars, people occupying territory on the Hunza River in Kashmir, were causing trouble to the Ruler of Kashmir by attacking Kashmir outposts and were subsequently informed that roads were going to be driven through their country. Anticipating resistance, the Hunza-Nagar Expedition was mounted in September 1891, consisting of the l/5th Gurkhas, two hundred strong, with a section of the Hazara Mountain Battery. This small force, of which Guy Boisragon was a member, was under orders to proceed from Abbottabad to Gilgit. By 14th October, the force commenced the ascent of the Tragbal and Burzil Passes, 11,800 and 13,500 feet respectively above sea level. Thick snow and freezing conditions impeded progress, and frostbite caused the column commander to be evacuated. Guy Boisragon was given command of the detachment.

From Gilgit, the detachment proceeded to Chalt where it was joined by some six Bengal Sappers, thirty rifles of the 20th Punjab Infantry and three Kashmir Infantry battalions. The force now became the Hunza-Nagar Field Force with a strength of 1130 rifles.

The Thams (rulers) of the Hunza and Nagar, continued to resist having roads through their territory and, after an ultimatum had been ignored, the Field Force advanced on the Tham fort of Nilt, some nine miles upstream from Chalt on the Hunza River.

The fort at Nilt stood on the extremity of a ledge which overhung the Nilt nullah and was protected by a precipice. Upstream of Nilt, on the right bank of the river, was Maiun, and facing it, on the same side of the river as Nilt, was the fortress of Thol. These became the later objectives of the Force.

On 2nd December 1891, the Force, with the l/5th Gurkhas under Guy Boisragon in the lead, wound its way towards the fort. The fort itself was substantial, with walls fourteen feet high and eight feet thick, built of stone and reinforced strongly with stout baulks of timber, its one entrance a gateway at the southwest corner.

Lieutenant Guy Boisragon, with his men of the l/5ᵗʰ Gurkhas, was ordered to take the fort by assault, after Captain Aylmer accompanying him with his Bengal Sappers, had blown-in the entrance gate. Under fire from the enemy, often at point blank range, Guy Boisragon, with a section of Gurkhas, reached the area of the main gate, and, under their covering fire, Captain Aylmer and Sapper Hazara Singh ran in, placed gun-cotton charges, and lit the fuse. Unfortunately the fuse was faulty and Captain Aylmer, although wounded and facing almost certain death, went back again to ignite it. This time there was an explosion and under cover of the clouds of dust the small party of l/5ᵗʰ Gurkhas carried the breach.

However, there were not enough of them to carry the fort, so Guy Boisragon went to collect reinforcements. Leaving Lieutenant Badcock to contest the gateway, he set off under heavy fire and returned to find Badcock wounded but the gateway still held. The l/5ᵗʰ Gurkhas then surged into the fort and the battle was over.

For their part in this action and the conspicuous gallantry they had displayed. Lieutenant Guy Boisragon and Captain Aylmer were each awarded the Victoria Cross. Lieutenant Badcock received the Distinguished Service Order, and no less than nine men of the l/5ᵗʰ Gurkhas were awarded the Indian Order of Merit.

Guy Boisragon was also Mentioned in Despatches for his part in the assault on Nilt Fort.

The citation covering the award of the Victoria Cross to Lieutenant Guy Boisragon was contained in the London Gazette of 12ᵗʰ July 1892 and read as follows:

"For his conspicuous bravery in the assault and capture of the Nilt Fort on 2ⁿᵈ December, 1891.
This Officer led the assault with dash and determination, and forced his way through difficult obstacles to the inner gate, when he returned for reinforcements, moving intrepidly to and fro under a heavy cross-fire until he had collected sufficient men to relieve the hardly pressed storming party and drive the enemy from the fort".

The campaign continued, through to the storming of the Thol cliffs on the 20ᵗʰ December 1891, where another Victoria Cross was awarded to an officer of the 5ᵗʰ Gurkhas, Lieutenant J Manners-Smith, and thence to the end of enemy resistance. Having made temporary arrangements for the administration of the country, the Field Force was disbanded and the l/5ᵗʰ Gurkhas returned from Gilgit, where it had been based since January 1892, to its home at Abbottabad in August of that year.

Over the next decade Guy Boisragon saw a good deal more of frontier warfare. He served in the Waziristan Expedition of 1894-95 and took part in the major operations on the Punjab Frontier, in Tirah and the Samana in 1897-98. After a series of staff and extra-regimental appointments he returned to regimental duty in 1910 as a Lieutenant Colonel and was appointed Commandant of the 1st Battalion 5ᵗʰ Gurkha Rifles (Frontier Force).

He proceeded with the battalion to Egypt in November 1914, where it became part of the "Canal Defence Force". In May 1915 the battalion sailed for Gallipoli and saw action against the Turks at Achi Baba. In this action, for which he was Mentioned in Despatches, Guy Boisragon was shot through the knee-cap and evacuated to hospital in England, where he was not declared fit for service until June 1917. He was promoted to Brevet Colonel on 3ʳᵈ June 1915 and Colonel in 1916, and also received the Order of the Nile, 4ᵗʰ Class.

He was employed at the War Office in London until the end of World War 1 and retired from the Army in 1920.

He died at Biarritz, France, on 14ᵗʰ July 1931, aged 66 and was buried in Kensal Green Cemetery, North London on 4ᵗʰ August 1931. His name is included in the Punjab Frontier Force Memorial that was originally in the Sanctum crypt of St. Luke's Church, Chelsea and has now been moved to the National Army Museum. He was not married.

LIEUTENANT J MANNERS-SMITH VC

Indian Staff Corps attached 5th Gurkha (Rifle) Regiment
later 5th Royal Gurkha Rifles (Frontier Force)

Date of Action	*Campaign*
20th December 1891	*Hunza 1891*

John Manners-Smith was born in Lahore on the 30th August 1864, the fifth son of the late Charles Manners-Smith FRCS, a Surgeon General in the Indian Medical Service. He was educated at Trinity College, Stratford-on-Avon, and King Edward VI School, Norwich, where he progressed at cricket but excelled at rugby, becoming Captain of the 1st XV. In 1881 he was placed 37th in the list of successful candidates for the Royal Military College, Sandhurst, whence he passed out 7th in 1883. Then followed two years service with The Norfolk Regiment before he transferred to the Indian Staff Corps and on 8th February 1886 was appointed as a Lieutenant to the 5th Goorkha Regiment (The Hazara Goorkha Battalion).

In 1887 he left the Regiment for the Political Department and in 1891 he was on special duty in Gilgit at the same time as the Hunza-Nagar Expedition was being assembled at Chalt, (see Lieutenant G H Boisragon VC, on page 24).

The Field Force included the 5th Gurkha (Rifle) Regiment and Lieutenant Manners-Smith joined the Field Force as a Political Officer.

During the storming of the fortress at Nilt on 2nd December 1891, Manners-Smith led the supporting force of Puniali Levies and the handful of rifles of the 20th Punjab Infantry on to the projecting spur adjacent to the fort, and thence gave close support to the assault by the l/5th Gurkhas when the gateway battle was joined and the fortress captured.

The Field Force had halted for the night at Nilt after the successes of 2nd December and then attempted to advance and secure the enemy stronghold of Maiun. The extremely difficult terrain, seemingly impregnable enemy positions and the impassable nature of tracks for mules, led the Force Commander to abandon his original intentions. The Force, therefore, remained stationary for seventeen days whilst endeavours were made to find ways and means of attacking and clearing the enemy from his strong-points. In the end it was not by Maiun that the advance was made. After exhaustive reconnaissance, night after night, by elements of the force, a Dogra sepoy, Nagdu, of The Kashmir Bodyguard Regiment, from a patient study of the ground, concluded that although extremely hazardous, a way existed, possibly for cragsmen, up the face of the Thol Cliffs. He had already succeeded at a second attempt, to scale to a point close to four enemy sangars. These were cleverly sited to dominate the ravine approaches and to provide mutual support to Maiun, and were strongly held. It was decided to attempt the storming of the sangars by this difficult approach.

The plan of attack was simple and was to be executed on the night of 20th December. Before moonrise at 10pm, Lieutenant Manners-Smith, accompanied by Lieutenant M A P Taylor of the 5th Gurkhas, was to lead out one hundred men and, under cover of darkness, reach a point on the floor of the Nilt Ravine immediately below the four enemy sangars explored by Sepoy Nagdu. At dawn the assault force, covered by support fire provided by selected marksmen in position on the spur above Nilt Fort, was to climb the Thol Cliffs and overcome the enemy positions. The 'support fire' party comprised men of the 5th Gurkhas, the 20th Punjab Infantry and the Kashmir battalions, divided into four groups (one to cover each sangar), together with two seven-pounder guns of the Hazara Mountain Battery.

At 7pm. Lieutenant Manners-Smith and his men started for the selected hiding place in the ravine below the sangar objectives.

Sounds of drumbeats were heard from the direction of Maiun but it soon became clear that the first important phase was undetected by the enemy. Before daylight, the covering troops left camp to take up position on the projecting spur and at 8am, when it was light enough to see clearly, fire was opened on the four sangars by the spur party. The climbers then left their temporary shelter and started the ascent. The cliff was so steep that they remained out of sight of both the occupants of the sangars and the enemy positions on the opposite side of the ravine. Their objective was at a height of more than twelve hundred feet above the bed of the ravine, and, watched anxiously by their comrades, they surmounted laboriously one difficult passage after another, until only four hundred feet remained to be negotiated. At this point, Manners-Smith and his advance party of fifty men of the 5th Gurkhas faltered and came to a standstill. Descending to the nala, they made a fresh attempt, this time reaching a point only sixty yards from the nearest enemy sangar before shouting and drumbeats from Maiun Fort alerted the enemy to their presence. Spurred on by the danger confronting them, the enemy left the shelter of their sangars to hurl down rocks upon the climbers.

Displaying great coolness and judgement, Manners-Smith worked forward gradually towards the objective. Seizing every chance that offered, he and the leading Gurkhas reached the relatively flat ground on which the sangars were sited. The first was rapidly surrounded, shots were fired through the opening in its rear, and a rush made for the interior. Among the first to enter was Harkia Thapa, Manners-Smith's orderly, whose conspicuous gallantry earned for him the Indian Order of Merit. A number of the enemy were killed before they could escape and, of those who fled, the fire of the covering detachment accounted for many more. With the arrival of more men from below, the remaining three sangar positions were cleared and no time was lost in dealing with the neighbouring breastworks on the mountain-side. The effect of this brilliant coup was to turn completely the enemy's left flank and to threaten his retreat. Realisation of this was almost immediate as long lines of fugitives were observed streaming up the valley as they quit the strongholds of Maiun and Thol. Orders were issued for the pursuit and, having left detachments to deal with Thol and Maiun, the Force, led by the men of the 5th Gurkhas, pushed rapidly up the valley and reached Pisan, some seven miles distant from Nilt. The action of the 20th December brought about the final collapse of enemy resistance. Remaining enemy strong-points were either surrendered or abandoned on the approach of our troops. One hundred of the enemy had been killed and one hundred and twenty captured, but more importantly the Hunza and Nagar, the hitherto unconquered brigands of the Kashmir frontier had been subdued.

Lieutenant John Manners-Smith received the coveted award of the Victoria Cross that was announced in the London Gazette of 12th July 1892 in the following terms:

"For his conspicuous bravery when leading the storming party at the attack and capture of the strong position occupied by the enemy near Nilt, in the Hunza-Nagar Country, on the 20th December, 1891.

The position was, owing to the nature of the country, an extremely strong one, and had barred the advance of the force for seventeen days. It was eventually forced by a small party of 50 rifles, with another of equal strength in support. The first of these parties was under the command of Lieutenant Smith, and it was entirely owing to his splendid leading, and the coolness, combined with dash, he displayed while doing so, that a success was obtained. For nearly four hours, on the face of a cliff which was almost precipitous, he steadily moved his handful of men from point to point, as the difficulties of the ground and showers of stones from above gave him an opportunity, and during the whole of this time he was in such a position as to be unable to defend himself from any attack the enemy might choose to make.

He was the first man to reach the summit, within a few yards of one of the enemy's sungars, which was immediately rushed, Lieutenant Smith pistolling the first man".

Manners-Smith then returned to political duties. He had previously accompanied Sir Mortimer Durand on a mission to Sikkim in 1888 and subsequently to Kabul in 1893. He took part in the Tirah Campaign of 1897-1898 with the 3rd Sikh Infantry and thereafter held political appointments in Baluchistan, Kashmir, Gilgit, Rajputana, Bhopal and, from 1906-1916, in Kathmandu, Nepal in which place he became the British Resident. His last appointment in 1919 was as agent to the Governor General and Chief Commissioner of Ajmer-Merwara (Rajputana). He was made a Companion of the Most Eminent Order of the Indian Empire in 1894, a Commander of the Royal Victorian Order in 1911, and was awarded the Izzat-i-Afghani for services in Afghanistan.

In 1896 he married Bertha Mabel, the elder daughter of the late Philip Latham.

A number of tributes have been written about Manners-Smith's personal qualities. Sir Mortimer Durand wrote of him, "Manners-Smith was as kindly and trustworthy as he was brave. One could depend on him in all circumstances, not only upon his courage, but also upon his unselfish devotion to duty. Loyalty was the mainspring of his character". Another wrote, "He was a man of the toughest physique, great hardihood, unbounded energy, and indomitable courage. He was a keen sportsman and devoted to big game shooting and the rough tribesmen of the high Himalayan valleys used to declare that he was the only Englishman who could surpass them on their own mountainsides".

Sir John Smyth said in his "Story of the Victoria Cross" (page 117), "There are four VC's I have known who literally did not know the meaning of fear. These were Carton de Wiart, with his one eye and one arm, the most shot about VC of them all, General Lord Freyberg, Field Marshal Lord Gort, and Jack Manners-Smith".

He was badly mauled by a panther in 1919 when shooting in Rajputana but from this he recovered. He returned to England in June of that year suffering from a wasting disease, died in a nursing home in London on 6th January 1920 aged 55 and was buried in Kensal Green Cemetery.

CAPTAIN W G WALKER VC

4th Gurkha Rifles
later 4th Prince of Wales's Own Gurkha Rifles

Date of Action	*Campaign*
22nd April 1903	*Somaliland*

William George Walker was born in Naini Tal, India, on 28th May 1863, the son of Deputy Surgeon General W Walker Indian Medical Service. He was educated at Haileybury and St John's College, Oxford, where he obtained an Arts degree and was commissioned into The Suffolk Regiment on 29th August 1885. He was promoted Lieutenant on 28th May 1887 and joined the 1st Battalion, 4th Goorkha Regiment in that year, seeing service with it in the second Miranzai Expedition 1891-92, and the Waziristan campaign 1894-95.

In 1903, Captain Walker was serving with the Bikanir Camel Corps (Bikanir Ganga Risala) in the Somaliland Field Force and was present at the actions at Daratoleh and Jidballi, where he was twice Mentioned in Despatches. It was after the action at Daratoleh and while serving with the Berbera Bohottle Flying Column, that Captain Walker was awarded the Victoria Cross. The citation in the London Gazette of 7th August 1903 read:

"During the return of Major Gough's column to Donop on the 22nd April last, after the action at Daratoleh, the rear-guard got considerably in rear of the column, owing to the thick bush, and to having to hold their ground while wounded men were being placed on camels. At this time Captain Bruce was shot through the body from a distance of about 20 yards, and fell on the path unable to move.

Captains Walker and Rolland, two men of the 2nd Battalion King's African Rifles, one Sikh and one Somali of the Camel Corps were with him when he fell.

In the meantime the column being unaware of what had happened were getting further away. Captain Rolland then ran back some 500 yards and returned with assistance to bring off Captain Bruce, while Captain Walker and the men remained with that Officer, endeavouring to keep off the enemy, who were all round in the thick bush. This they succeeded in doing, though not before Captain Bruce was hit a second time, and the Sikh wounded. But for the gallant conduct displayed by these Officers and men, Captain Bruce must have fallen into the hands of the enemy".

In 1904, W G Walker, now a Brevet Lieutenant Colonel, returned to the 1st Battalion, 4th Gurkha Rifles. In 1907 he was transferred to the 1st Gurkha Rifles and spent two years with that Regiment before re-joining the 4th in 1909. He was appointed Commandant of the 1st Battalion 4th Gurkha Rifles in 1910 having been promoted Colonel on 8th May of that year, after a short spell as Inspector of Imperial Service Troops, Bombay, sailed with the battalion for Suez on 26th August 1914 on the British India Steamship Company's ship "Baroda", becoming the first unit of the Indian Army to leave India for service in the Great War. The battalion then formed part of the Indian Expeditionary Force 'A'. This force was composed of two Divisions, the 3rd (Lahore) Division, and the 7th (Meerut) Division.

Soon after arrival in Egypt, Colonel Walker acted in command of the Sirhind Brigade of Lahore Division as the Brigadier had already gone to France, where the Brigade followed in 1914. He was confirmed as General Officer Commanding Sirhind Brigade in January 1915 and Brigadier General commanding both the Sirhind and Jullundur Brigades later the same year. He was promoted Major General in January 1916 and commanded the 2nd (British) Division in France. He retired in 1919, having been appointed a Companion of the Order of The Bath in 1914 and being Mentioned in Despatches three times.

On 23rd February 1907, he married in Melbourne, Australia, Margaret Alice Elaine Molesworth, daughter of Judge H Molesworth, by whom he had a son and daughter.

He died at Seaford, Sussex on 16th February 1936, aged 72 and was cremated at Woodvale Cemetery, Brighton on 19th February.

LIEUTENANT J D GRANT VC

8th Gurkha Rifles

Date of Action	*Campaign*
6th July 1904	*Tibet*

John Duncan Grant was born at Roorkee in the United Provinces, India on 28th December 1877, the eldest son of Colonel Suene Grant RE. He was educated at Cheltenham College and attended the Royal Military College, Sandhurst in 1898. He then joined the 30th Punjab Regiment of Bengal Infantry before transferring to the 44th Gurkha (Rifle) Regiment of Bengal Infantry, which later became the 1st Battalion, 8th Gurkha Rifles.

In the late 1890's, the Indian Government had become increasingly concerned about the behaviour of their neighbour Tibet. The Lhasa Government had been flirting with Russia and there were signs that a secret treaty with Moscow was being negotiated. This would have contravened Tibet's treaty with India. In 1903, Russian arms began to appear in the country, Indian territory was encroached upon, and it became clear that some protective measures had to be taken. A political officer was therefore sent to the frontier to treat with the Tibetans; the 8th Gurkha Rifles escorted him.

After waiting in vain for four months for the Tibetans to come to negotiate, the Indian Government reluctantly ordered an advance on Gyantse. At first the march was peaceful, but soon serious fighting took place, one action involving the 8th Gurkha Rifles being fought at 18,000 feet, probably a record height for an infantry battle at that time. One of the key positions in the Gyantse valley was Gyantse Jong, a fort built on a rock rising about 600 feet above the valley. It was held by about six thousand Tibetans.

The attack on Gyantse Jong ordered for 6th July 1904, was to be undertaken in two phases, the final phase being the capture of the fort by two companies of the 8th Gurkhas and one company of The Royal Fusiliers. After a preliminary artillery bombardment, Lieutenant Grant crossed the start line at 3.30pm and led his company up the precipitous slope. It was a bare rock face with little or no cover available, and the advance virtually had to be in single file to take advantage of firm footing. Showers of stones and heavy rifle fire met the company. Lieutenant Grant, and with him Havildar Karbir Pun, were hurled back more than once, but continued the advance. On reaching the top, Lieutenant Grant was wounded and thrown back once again, but despite his wounds he continued to climb, covered by the fire of his company below. Finally Grant and the Havildar, who had also been wounded, succeeded in forcing an entrance into the fortress. The other two companies followed up immediately and by 6pm Gyantse Jong had been captured. For his conspicuous gallantry on that day, Lieutenant Grant was awarded the Victoria Cross, while Havildar Karbir Pun received the Indian Order of Merit, 1st Class.

The citation in the London Gazette of 24th January 1905 read:

"On the occasion of the storming of the Gyantse Jong on 6th July 1904, the storming company, headed by Lieutenant Grant, on emerging from the cover of the village, had to advance up a bare, almost precipitous, rock-face, with little or no cover available, and under a heavy fire from the curtain, flanking towers on both sides of the curtain, and other buildings higher up the Jong. Showers of rocks and stones were at the time being hurled down the hillside by the enemy from above. One man could only go up at a time, crawling on hands and knees, to the breach in the curtain.

Lieutenant Grant, followed by Havildar Karbir Pun, 8th Gurkha Rifles, at once attempted to scale it, but on reaching the top he was wounded and hurled back, as was also the Havildar, who fell down the rock some 30 feet.

Regardless of their injuries, they again attempted to scale the breach, and, covered by the fire of the men below, were successful in their object, the Havildar shooting one of the enemy on gaining the top. The successful issue of the assault was very greatly due to the splendid example shown by Lieutenant Grant and Havildar Karbir Pun.

The latter has been recommended for the Indian Order of Merit".

In 1908 Captain Grant attended the Staff College in Quetta and was appointed Brigade Major, 5th (Jhelum) Brigade, 2nd (Rawalpindi) Division before becoming Brigade Major, 22nd (Lucknow) Brigade of the 8th (Lucknow) Division the following year. During the First World War, he saw service in 1917 as a Brigade Major in Egypt and Mesopotamia, where he was again wounded and Mentioned in Despatches. In June 1916 he became GSO I in Army Headquarters, New Delhi before being appointed an acting Lieutenant Colonel and Commandant of the 3rd Battalion 11th Gurkha Rifles on 4th June 1918.

After the war he served in Afghanistan in 1919, where he was again Mentioned in Despatches. In 1920 he was given temporary command of the 3/11 Gurkha Rifles in Waziristan for nine months before being appointed Commandant of the 1/10 Gurkha Rifles on 1st February 1921. For his service in Waziristan he was again Mentioned in Despatches, and awarded the DSO. He finished his active service in 1928 as a Brevet Colonel, having served for four years as Assistant Adjutant General in Army Headquarters and being appointed a Companion of the Order of The Bath on 3rd June 1929.

In 1934 he was appointed Colonel of the Regiment 10th Gurkha Rifles, an appointment he held for thirteen years. In 1907 he married Kathleen, the only daughter of Colonel Sir Peter Freyer KCB and they had one son and one daughter. He died in Tunbridge Wells, Kent on 20th December 1967, just before his 90th birthday.

Photograph of Gyantse Jong taken by Captain D W Humphreys the day after the Assault, which took place on 6th July 1904. Dotted line shows line of advance.

RIFLEMAN KULBIR THAPA VC

3rd Queen Alexandra's Own Gurkha Rifles

Date of Action	*Campaign*
25th/26th September 1915	*France*

Rifleman Kulbir Thapa was born at Nigalpani in the Palpa district of Nepal in about 1888. He was a Thapa of the Magar tribe, one of the tribes enlisted in seven of the ten Gurkha Regiments of the Indian Army before 1948, the 3rd Queen Alexandra's Own Gurkha Rifles being composed of men of the Magar and Gurung tribes in roughly equal proportions.

Kulbir Thapa originally enlisted in the 1st Battalion, 3rd Gurkha Rifles, but was transferred to the 2nd Battalion, for service in France in 1914.

In September 1915 the Indian Corps in France was required to carry out a holding attack in the neighbourhood of Loos near Neuve Chapelle. The task allotted to the 3rd Gurkhas in the Garhwal Brigade of the 7th Meerut Division was to act as the pivot on which the whole operation rested. Gas shells, which the British had proposed to use in certain circumstances, had been hit by mortar fire and the battalion had, therefore, to carry out the attack wearing gas masks. The Germans, on the other hand, were unharmed by the gas. They also realised that the British artillery bombardment had not cut their barbed wire defences as the British had hoped. As a result, the Germans began to mow down the attackers with machine-gun fire, the operation mounted by the Garhwal Brigade and the 3rd Gurkhas suffered severely.

Kulbir was in a party led by an officer which fought its way into a German trench and Kulbir, although wounded, was the sole survivor. For his exceptional courage in this action, Kulbir was awarded the Victoria Cross, the recommendation being most strongly supported by Officers of the 39th Garhwal Rifles and one Officer of the 2nd Battalion The Leicestershire Regiment.

The official citation for the award to Kulbir published in the London Gazette of 18th November 1915 read as follows:

His Majesty the KING has been graciously pleased to approve of the grant of the Victoria Cross to the undermentioned:
2129 Rifleman Kulbir Thapa, 2nd Battalion, 3rd Queen Alexandra's Own Gurkha Rifles.

"For most conspicuous bravery during operations against the German trenches south of Mauquissart.

When himself wounded, on the 25th September, 1915, he found a badly wounded soldier of the 2nd Leicestershire Regiment behind the first line German trench, and, though urged by the British soldier to save himself, he remained with him all day and night. In the early morning of the 26th September, in misty weather, he brought him out through the German wire, and, leaving him in a place of comparative safety, returned and brought in two wounded Gurkhas one after the other. He then went back in broad daylight for the British soldier and brought him in also, carrying him most of the way and being at most points under the enemy's fire".

Kulbir Thapa recovered from his wound, was promoted to Havildar, and proceeded to Egypt with his Regiment.

At the conclusion of the war he returned with his battalion to India and subsequent pension. He died in Nepal on 3rd October 1956 aged 68.

MAJOR G C WHEELER VC

9th Gurkha Rifles

Date of Action
23rd February 1917

Campaign
Mesopotamia

George Campbell Wheeler was born in Tokyo, Japan, on 7th April 1880, the son of Doctor Edwin Wheeler who had settled in Yokohama after being a Fleet Surgeon in the Royal Navy. He was educated at Bedford School from 1893-97 and played both cricket and rugby for the school. He was also an accomplished athlete. After leaving school, he entered the Royal Military College, Sandhurst, whence he was commissioned in 1900, joining the 9th Gurkha Rifles in 1901 and being promoted to Major in 1915.

In 1916, Wheeler's battalion, the 2nd Battalion 9th Gurkha Rifles, sailed for Mesopotamia to join 37th Indian Infantry Brigade of 7th Meerut Division. The 1st Battalion, 2nd King Edward's Own Gurkha Rifles was in the same Brigade.

The 7th Division was part of a force assembled to capture Baghdad. This involved 37 Brigade carrying out practice crossings of the River Tigris as a prelude to making an opposed crossing at a point where that river was some 360 yards wide.

On 23rd February 1917, Major Wheeler with 'D' Company of 2/9th Gurkha Rifles, crossed the river at Shumran in the face of very heavy enemy fire and established themselves on the far bank despite a determined counter-attack by the Turks. For his conduct on this day, Major Wheeler was awarded the Victoria Cross.

The official citation from the London Gazette of 8th June 1917, read as follows:

His Majesty the KING has been graciously pleased to approve the award of the Victoria Cross to the undermentioned officer: Maj. George Campbell Wheeler, Gurkha Rifles, Ind. A.

"For the most conspicuous bravery and determination.

This officer, together with one Gurkha officer and eight men, crossed a river and immediately rushed the enemy's trench under heavy bombing, rifle, machinegun, and artillery fire.

Having obtained a footing on the river bank, he was almost immediately afterwards counter-attacked by a strong enemy party with bombers. Major Campbell Wheeler at once led a charge with another officer and three men, receiving a severe bayonet wound in the head, but managed, in spite of this, to disperse the enemy.

This bold action on his part undoubtedly saved the situation. In spite of his wound, he continued to consolidate his position".

Major Wheeler recovered from his wound and later in 1917 was Mentioned in Despatches. Before retiring from the Army, Major Wheeler commanded the 1st Battalion, 9th Gurkha Rifles during operations against the Moplahs in South India in 1920/21 and became its permanent Commandant from 1923. In 1927 he became Commandant and District Superintendent of Police, Port Blair, Andaman Islands. He retired in 1930, died at Barton-on-Sea, Hampshire, England on 26th August 1938, aged 58 and was buried in the St. Magdalene Cemetery at New Milton.

RIFLEMAN KARANBAHADUR RANA VC

3rd Queen Alexandra's Own Gurkha Rifles

Date of Action	*Campaign*
10th April 1918	*Palestine*

Rifleman Karanbahadur was born at Mangalthan Gulmi, Litung, in the Baglung District of [Western] Nepal in 1898 and was a Rana of the Magar tribe. As a young soldier he was posted to the 2nd Battalion, 3rd Queen Alexandra's Own Gurkha Rifles as a reinforcement, the battalion then serving in Palestine as part of 232nd Infantry Brigade of 75th Infantry Division.

In April 1918, at El Kefr in Palestine, the battalion was engaged in an attack on a position held by the Germans on top of a rocky slope. The ground was broken and covered with scrub, which offered considerable tactical advantage to the enemy. The nature of the ground put 2nd Battalion 3rd Queen Alexandra's Own Gurkha Rifles in difficulties from the start of their advance because every movement was visible to the Germans.

The attack ended with heavy losses in 'B' Company, including the Company Commander but it was in this action that Karanbahadur earned his Victoria Cross.

The citation in the London Gazette of 21st June 1918 read:

His Majesty the KING has been graciously pleased to approve of the award of the Victoria Cross to:
4146 Riflmn Karanhahadur Rana. Gurkha Rif.

"For most conspicuous bravery, resource in action under adverse conditions, and utter contempt for danger.

During an attack he, with a few other men, succeeded under intense fire, in creeping forward with a Lewis gun in order to engage an enemy machine gun which had caused severe casualties to officers and other ranks who had attempted to put it out of action.

No 1 of the Lewis gun opened fire, and was shot immediately. Without a moment's hesitation Rifleman Karanbahadur pushed the dead man off the gun, and in spite of bombs thrown at him and heavy fire from both flanks, he opened fire and knocked out the enemy machine-gun crew; then, switching his fire on to the enemy bombers and riflemen in front of him, he silenced their fire. He kept his gun in action and showed the greatest coolness in removing defects which on two occasions prevented the gun from firing. During the remainder of the day he did magnificent work, and when a withdrawal was ordered he assisted with covering fire until the enemy were close on him. He displayed throughout a very high standard of valour and devotion to duty".

Rifleman Karanbahadur Rana received his Victoria Cross from King George V at Buckingham Palace in 1919. He died at Litung, Bharse Gulmi, Nepal on 25th July 1973 aged 74.

SUBEDAR LALBAHADUR THAPA VC

2nd King Edward VIIs Own Gurkha Rifles (The Sirmoor Rifles)

Date of Action	*Campaign*
5th/6th April 1943	*Tunisia*

Subedar Lalbahadur, a Thapa of the Magar tribe, was born in 1907 in the Baglung district of Western Nepal. He was enlisted into the 1st Battalion 2nd King Edward's Own Gurkha Rifles in 1925 and served with his battalion on the North West Frontier of India in 1936/37. In 1937 he was commissioned as a Gurkha Officer in the rank of Jemedar and promoted Subedar in 1940.

In 1941 the Battalion sailed for the Middle East and in 1942 was part of 7th Indian Infantry Brigade of the 4th Indian Division, whose objective was, as part of the Eighth Army, to oust the German and Italian forces from North Africa.

In Tunisia in April 1943, the plan was for the 4th Indian Division to take the Fatnassa heights in a silent night assault to enable the Eighth Army to wheel round behind Rommel's forces. The intention was that the l/2nd Gurkha Rifles and the l/9th Gurkha Rifles would capture the summit of the heights and they succeeded in doing so. In this action, the performance of Subedar Lalbahadur Thapa, second-in-command of "D" Company, was exemplary and he was recommended for an immediate MC. This was upgraded to the VC by the army commander and the citation that appeared in the London Gazette of 15th June 1943, read:

"On the night of 5th/6th April, 1943, during the silent attack on the Rass-Ez-Zouai feature, Subadar Lalbahadur Thapa was second in command of "D" Company. The Commander of No 16 Platoon was detached with one Section to secure an isolated feature on the left of the Company's objective. Subadar Lalbahadur Thapa took command of the remaining two Sections and led them forward towards the main feature on the outer ridge, in order to break through and secure the one and only passage by which the vital commanding feature could be seized to cover the penetration of the Division into the hills. On the capture of these hills the whole success of the Corps plan depended.

First contact with the enemy was made at the foot of a pathway winding up a narrow cleft. This steep cleft was thickly studded with a series of enemy posts, the inner of which contained an anti-tank gun and the remainder medium machine-guns. After passing through the narrow cleft, one emerges into a small arena with very steep sides, some 200 feet in height, and in places sheer cliff. Into this arena and down its sides numbers of automatic weapons were trained and mortar fire directed.

The garrison of the outer posts were all killed by Subadar Lalbahadur Thapa and his men by kukri or bayonet in the first rush and the enemy then opened very heavy fire straight down the narrow enclosed pathway and steep arena sides. Subadar Lalbahadur Thapa led his men on and fought his way up the narrow gully straight through the enemy's fire, with little room to manoeuvre, in the face of intense and sustained machine-gun concentrations and the liberal use of grenades by the enemy.

The next machine-gun posts were dealt with, Subadar Lalbahadur Thapa personally killing two men with his kukri and two more with his revolver. This Gurkha officer continued to fight his way up the narrow bullet-swept approaches to the crest. He and two Riflemen managed to reach the crest, where Subadar Lalbahadur Thapa killed another two men with his kukri, the Riflemen killed two more and the rest fled. Subadar Lalbahadur Thapa then secured the whole feature and covered his Company's advance up the defile.

This pathway was found to be the only practicable route up the precipitous ridge, and by securing it the Company was able to deploy and mop up all enemy opposition on their objective. This objective was an essential feature covering the further advance of the Brigade and of the Division, as well as the bridgehead over the anti-tank ditch.

There is no doubt that the capture of this objective was entirely due to this act of unsurpassed bravery by Subadar Lalbahadur Thapa and his small party in forcing their way up the steep gully, and up the cliffs of the arena under withering fire. The outstanding leadership, gallantry and complete disregard for his own safety shown by Subadar Lalbahadur Thapa were an example to the whole Company, and the ruthless determination of this Gurkha officer to reach his objective and kill his enemy had a decisive effect on the success of the whole operation".

Subedar Lalbahadur stayed with the battalion for the remainder of the campaign in North Africa and for a short time in Italy. In 1944 he was appointed Subedar Major of 5/2nd Gurkha Rifles on the North West Frontier of India and, in 1947, of 2 GR Wing of the combined 2 GR and 8 GR Regimental Centre. He retired on 6th June 1949 after his Regiment moved to Malaya to become part of the British Army and was later appointed Honorary Captain (GCO). Other awards made to him included the Order of British India (1st Class) and the Medal of the Order of The Star of Nepal. Four of his sons, Tejbahadur, Nandabahadur, Ranbahadur and Pahalsing served in the Regiment, the latter becoming a Queen's Gurkha Officer.

Honorary Captain (GCO) Lalbahadur Thapa VC OBI died at Paklihawa, Nepal, on 19th October 1968, aged 62.

HAVILDAR GAJE GHALE VC

5th Royal Gurkha Rifles (Frontier Force)

Date of Action	*Campaign*
27th May 1943	*Burma*

Havildar Gaje Ghale, a Gurung of the somewhat exclusive Ghale clan, was born in Barpak village in the Gorkha district of Nepal, probably in 1922. In 1935 he enlisted as a boy recruit and on completion of his training joined the 2nd Battalion 5th Royal Gurkha Rifles (Frontier Force). He served in Waziristan during the North West Frontier campaign of 1936-39 and from the outbreak of the Second World War until June 1942 he was an instructor at the Regimental Centre in Abbottabad.

The 2nd Battalion 5th Royal Gurkha Rifles was one of the Gurkha battalions in 48th Brigade of 17th Indian Division which came out of Burma in the first Burma campaign of the war in 1942. After re-equipping, it was ready again for battle in May 1943.

Between the 24th and 27th May 1943, a superior force of Japanese attempted an advance into the Chin Hills. The 2nd Battalion 5th Royal Gurkha Rifles was part of the opposing force. After two assaults against a Japanese position had failed, a third attempt was made by troops which included a platoon commanded by Havildar Gaje Ghale. He had not previously been under fire and his platoon consisted of young and inexperienced Gurkhas. The platoon was ordered to attack the Japanese position, the approach to which was along a ridge only five yards wide in some places. Whilst preparing to assault the enemy, Gaje Ghale was hit in the arm, chest and leg by shrapnel from a Japanese grenade. He ignored these wounds, rallied his platoon, and personally led assault after assault until the Japanese were finally driven off with heavy losses. Gaje then consolidated his position and only on a direct order from an officer did he report to the Regimental Aid Post for attention to his wounds.

The citation, published in the London Gazette of 30th September 1943 read:

The KING has been graciously pleased to approve the award of the VICTORIA CROSS to:
No 6816 Havildar Gaje Ghale, 5th Royal Gurkha Rifles (Frontier Force), Indian Army.

"In order to stop an advance into the Chin Hills of greatly superior Japanese forces it was essential to capture Basha East hill which was the key to the enemy position.

Two assaults had failed but a third assault was ordered to be carried out by two platoons of Havildar Gaje Ghale's company and two companies of another battalion.

Havildar Gaje Ghale was in command of one platoon: he had never been under fire before and the platoon consisted of young soldiers.

The approach for this platoon to their objective was along a narrow knife-edge with precipitous sides and bare of jungle whereas the enemy positions were well concealed. In places, the approach was no more than five yards wide and was covered by a dozen machine guns besides being subjected to artillery and mortar fire from the reverse slope of the hill.

While preparing for the attack the platoon came under heavy mortar fire but Havildar Gaje Ghale rallied them and led them forward.

Approaching to close range of the well-entrenched enemy, the platoon came under withering fire and this N.C.O. was wounded in the arm, chest and leg by an enemy hand grenade.

Without pausing to attend to his serious wounds and with no heed to the intensive fire from all sides, Havildar Gaje Ghale closed his men and led them to close grips with the enemy when a bitter hand to hand struggle ensued.

Havildar Gaje Ghale dominated the fight by his outstanding example of dauntless courage and superb leadership. Hurling hand grenades, covered in blood from his own neglected wounds, he led assault after assault encouraging his platoon by shouting the Gurkha's battle-cry*.

Spurred on by the irresistible will of their leader to win, the platoon stormed and carried the hill by a magnificent all out effort and inflicted very heavy casualties on the Japanese.

Havildar Gaje Ghale then held and consolidated this hard won position under heavy fire and it was not until the consolidation was well in hand that he went, refusing help, to the Regimental Aid Post, when ordered to do so by an officer.

The courage, determination and leadership of this N.C.O. under the most trying conditions were beyond all praise".

* *"Ayo Gurkhali" ("The Gurkhas are upon you").*

Gaje Ghale was promoted to Jemadar in 1943 and received his Victoria Cross from the Viceroy, Field Marshal Lord Wavell, at Delhi in 1944 and was later decorated by the Prime Minister of Nepal in Kathmandu with the Star of Nepal. He also represented his regiment on the Victory Parade in London in 1946.

When India gained Independence in 1947, Gaje Ghale continued to serve in his regiment, re-designated 5th Gorkha Rifles (Frontier Force) and was promoted Subedar in 1948. Later appointed Subedar Major of the 2nd Battalion he served in this capacity with the United Nations Force in the Congo in 1962-63. He was promoted to the honorary rank of Captain and on retirement settled in Almora, Uttar Pradesh in the Republic of India.

Regularly attending the Victoria Cross and George Cross Association reunions in the United Kingdom, he was presented to Her Majesty The Queen during her state visit to Nepal in February 1986.

On 19th August 1995, Gaje Ghale was present on the 'Last Big Parade' in London, the National event to commemorate the 50th Anniversary of the end of World War 2.

Honorary Captain Gaje Ghale VC died in New Delhi on 28th March 2000.

17 Ind. Div.

LIEUTENANT (ACTING CAPTAIN) M ALLMAND VC

6th Duke of Connaught's Own Lancers
attached 6th Gurkha Rifles
later 6th Queen Elizabeth's Own Gurkha Rifles

Date of Action   *Campaign*
11th/23rd June 1944   *Burma*

Michael Allmand was born in London on 22nd August 1923, the son of Professor A J Allmand MC, Professor of Chemistry, King's College, London. From preparatory school he went on to continue his education at Ampleforth College in Yorkshire, where he was regarded as having great personal charm and independent views. He had a deep and abiding religious faith, and in addition a clear idea of what the war was about, declaring that he would be proud to fight for the cause in which his country was involved.

From Ampleforth, Allmand went up to Oriel College, Oxford in 1941 to read History. After a year, he was called up for military service in the Royal Armoured Corps. In 1943 he was commissioned from the Officer Training School, Mhow into the Indian Armoured Corps, and joined the 6th Duke of Connaught's Own Lancers, a cavalry regiment of the Indian Army.

Whilst serving as an instructor at the I A C Armoured Car School Ferozepore, Allmand asked that he should be posted to a unit on active service. This was turned down, but shortly afterwards GHQ India called for officer volunteers to serve in the Second Chindit Expedition. Without delay, Allmand applied and was accepted for service in the Expedition. He soon found himself posted to the 3rd Battalion, 6th Gurkha Rifles, one of the battalions in 77th Indian Infantry Brigade. The Commander of that Brigade was the redoubtable Brigadier J M (Mad Mike) Calvert. The operations in which 77 Brigade were involved were intended to capture MOGAUNG. The atrocious monsoon weather conditions, and the swampy terrain in which they had to fight were appalling. As a result weary and exhausted soldiers succumbed quickly to sickness and trench foot.

Over a period, Allmand took part in a number of actions in all of which he proved himself to be a capable and cool-headed leader, despite the physical conditions in which he and his men had to function.

The official citation for the award of the Victoria Cross to Captain Allmand published in the London Gazette of 26th October 1944 read as follows:

The KING has been graciously pleased to approve the posthumous award of the VICTORIA CROSS to: Lieutenant (acting Captain) Michael Allmand (E.C. 8188), Indian Armoured Corps (attd. 6th Gurkha Rifles).

"Captain Allmand was commanding the leading platoon of a Company of the 6th Gurkha Rifles in Burma on 11th June, 1944, when the Battalion was ordered to attack the Pin Hmi Road Bridge.

The enemy had already succeeded in holding up our advance at this point for twenty-four hours. The approach to the Bridge was very narrow as the road was banked up and the low-lying land on either side was swampy and densely covered in jungle. The Japanese who were dug in along the banks of the road and in the jungle with machine guns and small arms, were putting up the most desperate resistance.

As the platoon came within twenty yards of the Bridge, the enemy opened heavy and accurate fire, inflicting severe casualties and forcing the men to seek cover. Captain Allmand, however, with the utmost gallantry charged on by himself, hurling grenades into the enemy gun positions and killing three Japanese himself with his kukrie.

Inspired by the splendid example of their platoon commander the surviving men followed him and captured their objective.

Two days later Captain Allmand, owing to casualties among the officers, took over command of the Company and, dashing thirty yards ahead of it through long grass and marshy ground, swept by machine gun fire, personally killed a number of enemy machine gunners and successfully led his men onto the ridge of high ground that they had been ordered to seize.

Once again on June 23rd in the final attack on the Railway Bridge at Mogaung, Captain Allmand, although suffering from trench-foot, which made it difficult for him to walk, moved forward alone through deep mud and shell-holes and charged a Japanese machine gun nest single-handed, but he was mortally wounded and died shortly afterwards.

The superb gallantry, outstanding leadership and protracted heroism of this very brave officer were a wonderful example to the whole Battalion and in the highest traditions of his regiment".

Captain Allmand was just 20 when he was killed. At an impressive and moving ceremonial parade held at Cassino Lines, Hong Kong, on 22nd July 1991, the Allmand family presented Michael's Victoria Cross and campaign medals to the 6th Queen Elizabeth's Own Gurkha Rifles.

Chindits

RIFLEMAN GANJU LAMA VC, MM

7th Gurkha Rifles
later 7th Duke of Edinburgh's Own Gurkha Rifles

Date of Action	*Campaign*
12th June 1944	*Burma*

Ganju Lama was born at Sangmo Busty, Sikkim in 1924 and although he was enlisted into the 7th Gurkha Rifles in 1942, he was not an ethnic Gurkha nor was he a Nepalese subject. At the time of his enlistment during the Second World War, Gurkha Regiments were prepared to accept any potential recruits who closely resembled the Gurkha and who lived near the borders of Nepal. Ganju came from the Kingdom of Sikkim. At the time he was enlisted, Ganju's real name was "Gyamtso" which a clerk in the Recruiting Office wrote down as "Ganju" and Ganju he remained.

After leaving the Regimental Centre in 1943 Ganju joined the 1st Battalion 7th Gurkha Rifles in Burma near Imphal. In May 1944 during operations on the Tiddim Road 1/7 GR surprised a party of Japanese and killed several of them. For his part in this action Ganju was awarded the Military Medal.

By June 1944 the 14th Army in Burma had made steady progress in clearing the Japanese from the area round Imphal and Kohima, although the Japanese were not yet on the run. One morning the enemy launched a major attack, with tanks in support, on the 2nd Battalion 5th Royal Gurkha Rifles, which was in the same Brigade (48th Indian Infantry Brigade, 17th Indian Division) as 1/7 GR. The action was not unduly prolonged thanks to a counter-attack by 1/7 GR in order to relieve the pressure on the 2/5th.

In this action it was Ganju Lama's skilful use of the PIAT that resulted in the destruction of two enemy tanks but he was himself seriously wounded. For this action he was awarded the Victoria Cross.

The citation published in the London Gazette of 7th September 1944 read:

"In Burma, on the morning of the 12th June, 1944, the enemy put down an intense artillery barrage lasting an hour on our positions north of the village of Ningthoukhong. This heavy artillery fire knocked out several bunkers and caused heavy casualties, and was immediately followed by a very strong enemy attack supported by five medium tanks. After fierce hand to hand fighting, the perimeter was driven in one place and enemy infantry, supported by three medium tanks, broke through, pinning our troops to the ground with intense fire.

"B" Company, 7th Gurkha Rifles, was ordered to counter-attack and restore the situation. Shortly after passing the starting line it came under heavy enemy medium machine-gun and tank machine-gun fire at point blank range, which covered all lines of approach. Rifleman Ganju Lama, the No. 1 of the P.I.A.T. gun, on his own initiative, with great coolness and complete disregard for his own safety, crawled forward and engaged the tanks single handed. In spite of a broken left wrist and two other wounds, one in his right hand and one in his leg, caused by withering cross fire concentrated upon him, Rifleman Ganju Lama succeeded in bringing his gun into action within thirty yards of the enemy tanks and knocked out first one and then another, the third tank being destroyed by an anti-tank gun.

In spite of his serious wounds, he then moved forward and engaged with grenades the tank crews, who now attempted to escape. Not until he had killed or wounded them all, thus enabling his company to push forward, did he allow himself to be taken back to the Regimental Aid Post to have his wounds dressed.

Throughout this action Rifleman Ganju Lama, although very seriously wounded, showed a complete disregard for his own personal safety, outstanding devotion to duty and a determination to destroy the enemy which was an example and an inspiration to all ranks. It was solely due to his prompt action and brave conduct that a most critical situation was averted, all positions regained and very heavy casualties inflicted on the enemy".

After this exploit Ganju Lama was rescued by men of his own section, and taken by stretcher to the Regimental Aid Post prior to being evacuated to a Base Hospital. Ganju Lama was eventually traced through the chain of evacuation and was presented with his Victoria Cross in Delhi by His Excellency the Viceroy, Field Marshal Lord Wavell, in the presence of Admiral Lord Louis Mountbatten, General Slim, and members of his own family from Sikkim.

In 1946 The Maharaja of Sikkim, (then an independent kingdom) declared a National holiday in Ganju's honour, made a gift of land and later appointed him *'Pema Dorje'* (PD).

After Indian Independence in 1947 Ganju Lama joined the 11th Gorkha Rifles, a re-raised Regiment comprising Gurkhas of the 7th and 10th Gurkha Rifles who had opted to continue their service with India as opposed to joining the British Army. In due course he was promoted Subedar Major and in 1965 appointed ADC to the President of India. He was granted the honorary rank of Captain in 1968 while still serving and, on retirement, returned to his home in Sikkim having been appointed Honorary ADC to the President for life.

Regularly attending the Victoria Cross and George Cross Association reunions in the United Kingdom and a number of Regimental celebrations worldwide, he was presented to Her Majesty The Queen during her state visit to Nepal in February 1986.

The programme for the 1995 Royal Tournament in London included an arena display re-enactment of Ganju's VC action. On 24th July, at the end of the re-enactment and in the presence of Her Majesty The Queen, Ganju was brought into the arena in an open Rolls Royce car, spotlighted and introduced to the audience; he was received with tremendous acclaim!

On 19th and 20th August 1995, Ganju Lama took part in the United Kingdom National events in London to commemorate the 50th Anniversary of the end of World War 2.

Honorary Captain Ganju Lama VC MM PD died at his home in Sikkim on 1st July 2000. His funeral on 5th July was attended by senior Ministers and officials of the State Government, a large contingent of the Indian Army led by the GOC of the Sikkim based 17 Division and representatives from many organisations, including the British High Commission New Delhi and Regimental Associations. Some 5000 people were in the funeral procession, a dignified and moving event honouring a National hero.

CLEARING A ROAD BLOCK

RIFLEMAN TULBAHADUR PUN VC

6th Gurkha Rifles
later 6th Queen Elizabeth's Own Gurkha Rifles

Date of Action	*Campaign*
23rd June 1944	*Burma*

Rifleman Tulbahadur Pun who was born in 1923 was, like so many recruits enlisted into the Gurkha Brigade from Western Nepal, recruited from a hill village called Banduk, in the Gulmi Tehsil of 4,000 Parbat District, west of Kathmandu. He enlisted in 1941 and after completing his recruit training in the Regimental Centre of the 6th Gurkha Rifles at Abbottabad in Northern India (now Pakistan), he joined the 3rd Battalion which was part of the Chindit Force in Burma.

On the 23rd June 1944, the battalion was involved in an action in which Rifleman Tulbahadur Pun distinguished himself and earned a Victoria Cross.

The citation in the London Gazette dated 9th November 1944 read:

The KING has been graciously pleased to approve the award of the VICTORIA CROSS to:
No. 10119 Rifleman Tulbahadur Pun, 6th Gurkha Rifles, Indian Army.

"In Burma on June 23rd 1944, a Battalion of the 6th Gurkha Rifles was ordered to attack the Railway Bridge at Mogaung. Immediately the attack developed the enemy opened concentrated and sustained cross fire at close range from a position known as the Red House and from a strong bunker position two hundred yards to the left of it.

So intense was this cross fire that both the leading platoons of "B" Company, one of which was Rifleman Tulbahadur Pun's, were pinned to the ground and the whole of his Section was wiped out with the exception of himself, the Section Commander and one other man. The Section Commander immediately led the remaining two men in a charge on the Red House but was at once badly wounded. Rifleman Tulbahadur Pun and his remaining companion continued the charge, but the latter too was immediately badly wounded.

Rifleman Tulbahadur Pun then seized the Bren Gun, and firing from the hip as he went, continued the charge on this heavily bunkered position alone, in the face of the most shattering concentration of automatic fire, directed straight at him. With the dawn coming up behind him, he presented a perfect target to the Japanese. He had to move for thirty yards over open ground, ankle deep in mud, through shell holes and over fallen trees.

Despite these overwhelming odds, he reached the Red House and closed with the Japanese occupants. He killed three and put five more to flight and captured two light machine guns and much ammunition. He then gave accurate supporting fire from the bunker to the remainder of his platoon which enabled them to reach their objective.

His outstanding courage and superb gallantry in the face of odds which meant almost certain death were inspiring to all ranks and were beyond praise".

Rifleman Tulbahadur Pun received his Victoria Cross from His Excellency the Viceroy, Field Marshal Lord Wavell, at a special parade held in Delhi on 3rd March 1945.

After Indian Independence in 1947 his regiment was transferred to the British army; Tulbahadur Pun joined the 2nd Battalion

6th Gurkha Rifles in The Brigade of Gurkhas and served in Malaya [Malaysia] and Hong Kong. Throughout his service he continued to display operational efficiency and courage that were an inspiration to his younger companions and especially during the Malayan Emergency [1948 – 1960].

He rose to Warrant Officer rank and was appointed Regimental Sergeant Major of his battalion. On retirement on 14th May 1959 he was promoted to Honorary Lieutenant (QGO) and returned to his village in Nepal.

In retirement he visited the United Kingdom to attend Victoria Cross and George Cross Association reunions and also visited his Regiment when stationed in the Far East. He was presented to Her Majesty The Queen during her state visit to Nepal in February 1986. He died in Nepal on 20th April 2011.

Chindits

JEMEDAR (ACTING SUBEDAR) NETRABAHADUR THAPA VC

5th Royal Gurkha Rifles (Frontier Force)

Date of Action	*Campaign*
25th/26th June 1944	*Burma*

Netrabahadur Thapa was born in 1916 at Rahu in the Lamjung district of Nepal and enlisted in the 5th Royal Gurkha Rifles (Frontier Force) on 8th January 1932. After training he joined the 2nd Battalion of his Regiment as a Rifleman. Subsequently promoted to Lance Naik and then to Naik, he served on the North West Frontier in 1936/37 before joining the recruit training staff of the Regimental Depot at Abbottabad in 1940 and was promoted to Havildar. Because of the severe casualties suffered by his Battalion in Burma, he, and his colleague, Havildar Gaje Ghale, who was also awarded the Victoria Cross, were posted to 2/5th Royal Gurkha Rifles in early 1942. On 27th December 1942 Netrabahadur was promoted Jemedar in command of a platoon.

In March 1944, 2/5th RGR was part of 17th Indian Division in direct contact with the enemy in the Chin Hills area, having been ordered to withdraw on Imphal to withstand the severe threat posed by the Japanese advance towards the Assam plains and Bengal.

Casualties mounted during the blocking and clearing operations that ensued along the Tiddim-Imphal road. Japanese determination was extreme and the commander of their 15th Army was resolved to place all of his resources into a final effort to take Imphal.

So began the final and bitter battle.

The 2nd Battalion 5th Royal Gurkha Rifles had been assigned to the Bishenpur area at the beginning of June by which time Netrabahadur Thapa had been granted the acting rank of Subedar.

The citation in the London Gazette of 12th October 1944, read as follows:

The KING has been graciously pleased to approve the posthumous award of the VICTORIA CROSS to:
Jemadar (acting Subadar) Netrabadadur Thapa (28467 IO), 5th Royal Gurkha Rifles (Frontier Force), Indian Army.

"Subadar Netrabadadur Thapa was in command of the garrison of 41 men of the 2/5th Royal Gurkha Rifles (Frontier Force) which on the afternoon of 25th June, 1944, took over the isolated piquet known as Mortar Bluff situated on the hillside commanding the base at Bishenpur in Burma. The piquet position, completely devoid of any cover, was situated some 400 yards from the next piquet, from which it could be supported to some extent by 3 inch mortar fire, but was commanded by Water Piquet, a short distance away on high ground to the South, which had been over-run by strong enemy forces on the previous night and was still in enemy hands. Owing to its commanding position the retention of Mortar Bluff was vital to the safety of other positions farther down the ridge and to Bishenpur itself.

The relief had been harassed by enemy snipers at close range but was completed at 1830 hours without casualties. A little more than an hour later the enemy began to attack. For this purpose a 75 millimetre and a 37 millimetre gun were bought on up to the high ground overlooking the position and poured shell after shell at point blank range for ten minutes into the narrow confines of the piquet, and this was followed by a determined attack by not less than one company of Japanese. A fierce fight ensued in which Subadar Netrabahadur Thapa's men, exhorted by their leader, held their ground against heavy odds and drove the enemy back with disproportionate losses. During this time Subadar Netrabadadur Thapa with tireless energy and contempt for his own safety moved from post to post encouraging his young N.C.O.s and riflemen, of which the garrison was largely composed, and tending the wounded.

A short lull followed during which Subadar Netrabahadur Thapa gave a clear and concise report on the telephone to his Commanding Officer and asked for more artillery defensive fire. Having done this he made preparations to meet the next onslaught which was not long in coming.

Under cover of the pitch dark night and torrential rain the enemy had moved round to the jungle from the cover of which they launched their next attack. Still in considerable strength and as determined and ferocious as ever the enemy poured out from the jungle across the short space of open ground to the piquet defences under cover of small arms and 37 millimetre gun fire from a flank. For a time our men held their ground until, as ill-luck would have it, both the L.M.G. and T.M.G. of one section jammed.

With much reduced fire-power the section were unable to hold on, and the enemy forced an entrance and over-ran this and another section, killing or wounding 12 out of the 16 men comprising the two sections. Having no reserve Subadar Netrabahadur Thapa himself went forward from his Headquarters and stemmed any further advance with grenades.

The situation was however critical. With more than half his men casualties, ammunition low, and the enemy in possession of part of his perimeter, Subadar Netrabahadur Thapa would have been justified in withdrawing, but in his next report to his Commanding Officer he stated that he intended holding on and asked for reinforcements and more ammunition.

So efficient were his plans for defence and such was the fine example of this gallant Gurkha officer that not a man moved from his trench and not a yard more ground was gained by the enemy, despite their desperate attempts.

Thus the night passed until at 0400 hours a section of 8 men with grenades and small arms ammunition arrived. Their arrival inevitably drew fire and all the 8 were soon casualties. Undismayed, however, Subadar Netrabahadur Thapa retrieved the ammunition and himself with his platoon Headquarters took the offensive armed with grenades and khukris. Whilst so doing he received a bullet wound in the mouth followed shortly afterwards by a grenade which killed him outright. His body was found next day, khukri in hand and a dead Japanese with a cleft skull by his side.

True to the traditions of the service and his race Subadar Netrabahadur Thapa had fought against overwhelming odds for 8 hours before he was killed. His fine example of personal bravery and his high sense of duty so inspired his men that a vital position was held to the limit of human endurance.

His valour and devotion to duty will remain an epic in the history of the Regiment".

The Victoria Cross was presented to his young widow, Nainasara Magarni, by His Excellency the Viceroy, Field Marshal Lord Wavell, on 23rd January 1945 at a special parade of the battalion at Nowshera in Northern India (now Pakistan).

17 Ind. Div.

RIFLEMAN (ACTING NAIK) AGANSING RAI VC

5th Royal Gurkha Rifles (Frontier Force)

Date of Action	*Campaign*
26th June 1944	*Burma*

Agansing Rai was born in Amsara, Chisankhu, in the Okhaldhunga District of Nepal in 1920. He was recruited into the 5th Royal Gurkha Rifles (Frontier Force) during 1941 and on completion of his training, joined the 2nd Battalion of the Regiment. He was made a Section Commander with the rank of Naik in 1943/44 and had already experienced action against the Japanese in the Chin Hills area of Burma during the first few months of 1944. In June of that year his Battalion, which formed part of 17th Indian Division, was under great pressure endeavouring to stem the fanatical Japanese assault on Imphal by their 15th Army. Its area of operations was Bishenpur, in particular the Bishenpur-Silchar track, which ran through the hills some four miles west of Bishenpur and which had been for some three months the scene of hard and bitter fighting.

On the morning of the 26th June 1944 "C" Company of 2/5th Royal Gurkha Rifles (Frontier Force) was ordered to recapture an enemy position dominating the Bishenpur-Silchar track.

The enemy position consisted of two piquets, one, "Water Piquet" overlooked the other, "Mortar Bluff", from a distance of two hundred yards and they were mutually supporting. Both piquets had been the scene of fierce fighting and had changed hands several times, "Mortar Bluff" having been captured the previous night (after the magnificent stand of Subedar Netrabahadur Thapa of the same Battalion, who was killed, but awarded a posthumous VC, (see page 56), as also was "Water Piquet", by greatly superior enemy forces. The possession of these two positions was vital to our communications up the Silchar track and enemy occupation of them for any length of time would have made our positions down the ridge towards Bishenpur untenable.

To the west of both piquet positions was dense jungle up to within a few yards, but the hillside was completely bare and open on the three other sides. The approach from any of these was up a slippery precipitous ridge to a false crest over which any assault had to be launched in full view of the enemy for a distance of eight hundred yards.

For his action on this day Naik Agansing Rai was awarded the Victoria Cross. The citation in the London Gazette dated 5th October 1944 read:

"In Burma on 24th and 25th June, 1944, after fierce fighting, the enemy, with greatly superior forces, had captured two posts known as "Water Piquet"" and "Mortar Bluff". These posts were well sighted and were mutually supporting and their possession by the enemy threatened our communications.

On the morning of 26th June, 1944, a Company of the 5th Royal Gurkha Rifles (Frontier Force), was ordered to recapture these positions.

After a preliminary artillery concentration, the Company went into the attack but on reaching a false crest about 80 yards from its objective, it was pinned down by heavy and accurate fire from a machine-gun in "Mortar Bluff" and a 37 millimetre gun in the jungle, suffering many casualties. Naik Agansing Rai. appreciating that more delay would inevitably result in heavier casualties, at once led his section under withering fire directly at the machine-gun and, firing as he went, charged the position, himself killing three of the crew of four. Inspired by this cool act of bravery the section surged forward across the bullet swept ground and routed the whole garrison of "Mortar Bluff".

This position was now under intense fire from the 37 millimetre gun in the jungle and from "Water Piquet". Naik Agansing

Rai at once advanced towards the gun, his section without hesitation following their gallant leader. Intense fire reduced the section to three men before half the distance had been covered but they pressed on to their objective. Arriving at close range, Naik Agansing Rai killed three of the crew and his men killed the other two. The party then returned to "Mortar Bluff" where the rest of their platoon were forming up for the final assault on "Water Piquet". In the subsequent advance heavy machine-gun fire and showers of grenades from an isolated bunker position caused further casualties. Once more, with indomitable courage, Naik Agansing Rai, covered by his Bren gunner, advanced alone with a grenade in one hand and his Thompson Sub-Machine gun in the other. Through devastating fire he reached the enemy position and with his grenade and bursts from his Thompson Sub-Machine gun killed all four occupants of the bunker.

The enemy, demoralised by this NCO's calm display of courage and complete contempt for danger, now fled before the onslaught on "Water Piquet" and this position too was captured.

Naik Agansing Rai's magnificent display of initiative, outstanding bravery and gallant leadership, so inspired the rest of the Company that, in spite of heavy casualties, the result of this important action was never in doubt".

The Victoria Cross was presented to Agansing Rai by His Excellency the Viceroy, Field Marshal Lord Wavell at a special parade of the Battalion at Nowshera on 23rd January 1945. After the presentation, having passed through a heavy barrage of cameras, Agansing defeated the Press reporters by his disarming smile and deprecating reply to their many requests for his feelings and reactions during the battle – "I'm sorry, I forget".

Agansing was posted to the Regimental Centre as an instructor, and together with Havildar Gaje Ghale VC (see page 46) took part in the Victory Parade in London in 1946.

He rejoined the 2nd Battalion 5th Royal Gurkha Rifles (Frontier Force) in Japan in 1947 and attained the rank of Subedar.

After Indian Independence in 1947, he remained with his Regiment in India and was appointed Subedar Major serving with his Battalion in the Congo in 1962/63 as part of the United Nations Force.

On retirement from the army he was granted the honorary rank of Lieutenant and returned to his village in Nepal.

He was one of the Gurkha VCs presented to Her Majesty the Queen in Kathmandu during her state visit to Nepal in February 1986. Always a man of stature and presence, he attended a number of Victoria Cross and George Cross Association reunions in London. In 1995 Agansing took part in the United Kingdom National events to commemorate the 50th Anniversary of the end of World War 2; on 19th August he marched down the The Mall at the head of the 'Last Big Parade'.

Honorary Lieutenant Agansing Rai VC died in Kathmandu, Nepal on 27th May 2000. His medals are displayed in the Ashcroft Gallery at the Imperial War Museum.

17 Ind. Div.

TAKAW FERRY—SALWEEN RIVER

CAPTAIN (TEMPORARY MAJOR) F G BLAKER VC, MC

The Highland Light Infantry (City of Glasgow Regiment)
attached 9th Gurkha Rifles

Date of Action	*Campaign*
9th July 1944	*Burma*

Frank Gerald Blaker was born at Kasauli, Punjab, India, on 8th May 1920, the son of Captain Blaker, who had served in Mesopotamia in World War 1 as a surgeon in the Royal Army Medical Corps. At the end of the war, Captain Blaker joined the Indian Medical Department and served as a civilian officer at the Kasauli Pasteur Institute in 1920, the Rangoon Pasteur Institute from 1921-26, Mandalay General Hospital from 1926-27, and as Director of the Vaccine Institute Meiktila from 1928-40. In 1940 Captain Blaker retired and the family moved to Kenya. When his father was stationed in Burma, Frank Blaker attended a number of schools, including the American Methodist School in Mandalay, St Paul's School, Darjeeling, and the Government English High School, Maymyo.

By the mid 1930s he was already a natural and accomplished sportsman, interested in the jungle and wildlife and fond of shooting. From 1937 to 1939 Frank Blaker was a boarder at Taunton School before joining The Somerset Light Infantry in March 1940, followed in 1941, by a commission in The Highland Light Infantry. At the end of 1941, he left the UK for India and in May 1942 joined the 3rd Battalion, 9th Gurkha Rifles, which then formed part of the 4th Indian Infantry Brigade of 26th Indian Division. The Division was operating in the Arakan with the object of containing any Japanese advance into India from that quarter. In July 1943 when the Battalion was based at Taung Bazaar, Blaker, then commanding "C" Company, was sent to investigate Japanese activity in a village five miles to the south. During the ensuing engagement and at the end of a two-mile chase, sixteen Japanese had been killed and three wounded prisoners taken. The latter included the first Japanese officer to be captured in the Arakan. For this action Captain Blaker was awarded the Military Cross.

On 5th July 1944, 3/9 GR marched along the Salman Chaung under the western slopes of Hill 2171, a jungle-covered spur which stood above Taunghi. Here the Japanese were determined to make a stand. By 6th July a path had been cleared to the summit and on 9th July, "B" and "C" Companies were entrusted with the task of taking the crest of the hill. The plan was that "B" Company would attack frontally up the southern face of the feature and "C" Company, under command of Major Blaker, would encircle the spur and attack from the other side. It was the intention that "B" Company would commence their attack when firing was heard from the other side of the spur. Unfortunately, "B" Company closed up on the Japanese outposts too soon and lost thirty men in a short space of time. Meanwhile, "C" Company, who had heard the firing, quickened their pace and almost at the top of Hill 2171 ran into heavy Japanese machine-gun fire. Having gone to ground and without room to manoeuvre, Blaker sprang to his feet and shouted for his men to charge. As he charged, Blaker fired at the machine-gun and received a grenade wound in his left wrist. When he was seven yards from the gun, he was fatally wounded by a long burst of fire. The men following pressed on, some Japanese were cut down and the rest fled. Resistance on Hill 2171 then weakened, enabling the front and rear companies to link up and the hill to be captured.

Captain Sweetman, who had taken over the Company when Blaker was wounded, found him 40 yards in rear with bullet holes in his stomach, chest and shoulder. Blaker, who knew he was going to die, ordered Sweetman to return to "C" Company but the latter ordered the Havildar Major to stay with Blaker. Soon after, Blaker told the Havildar Major that he was dying and asked that the names of two men he wished to be recommended for gallantry should be passed to Captain Sweetman. His last words, relayed to Captain Sweetman by the Havildar Major were "Thank "C" Company for all they have done for me. Tell them that I have gone from them, but they must go on fighting to the end."

The London Gazette dated 26th September 1944 announcing the award of a posthumous Victoria Cross read as follows:

The KING has been graciously pleased to approve the posthumous award of the VICTORIA CROSS to:

Captain (temporary Major) Frank Gerald Blaker. M.C. (193864), The Highland Light Infantry (City of Glasgow Regiment) (attached 9th Gurkha Rifles, Indian Army).

"In Burma on 9th July, 1944, a Company of the 9th Gurkha Rifles was ordered to carry out a wide, encircling movement across unknown and precipitous country, through dense jungle, to attack a strong enemy position on the summit of an important hill overlooking Taungni.

Major Blaker carried out this movement with the utmost precision and look up a position with his Company on the extreme right flank of the enemy, in itself a feat of considerable military skill.

Another Company, after bitter fighting, had succeeded in taking the forward edge of the enemy position by a frontal assault, but had failed to reach the main crest of the hill in the face of fierce opposition.

At this crucial moment Major Blaker's Company came under heavy and accurate fire at close range from a medium machine gun and two light machine guns, and their advance was also completely stopped.

Major Blaker then advanced ahead of his men through very heavy fire and, in spite of being severely wounded in the arm by a grenade, he located the machine guns, which were the pivot of the enemy defence, and single handed charged the position. When hit by a burst of three rounds through the body, he continued to cheer on his men while lying on the ground.

His fearless leadership and outstanding courage so inspired his Company that they stormed the hill and captured the objective while the enemy fled in terror into the jungle.

Major Blaker died of wounds while being evacuated from the battlefield. His heroism and self sacrifice were beyond all praise and contributed in no small way to the defeat of the enemy and the successful outcome of the operations."

Major Blaker is buried at the Taukkyan Military Cemetery on the outskirts of Rangoon, Burma.

A BASHA

no picture available

RIFLEMAN SHERBAHADUR THAPA VC

9ᵗʰ Gurkha Rifles

Date of Action	*Campaign*
18ᵗʰ/19ᵗʰ September 1944	*Italy*

Rifleman Sherbahadur was born at Ghalechap in the Tanhu district of Nepal in 1921 and was a Thapa of the Chettri tribe, one of the tribes enlisted by the 9ᵗʰ Gurkhas. He enlisted on the 20ᵗʰ November 1942 and after completing his training at the Regimental Centre was sent to the 1ˢᵗ Battalion 9ᵗʰ Gurkha Rifles in Italy with a draft to replace heavy casualties suffered during the second assault on Monte Cassino.

In September 1944 the Germans were holding a defensive position stretching east to west in northern Italy and known as the Gothic Line. This was regarded as the final defensive position for the protection of the "under belly of Europe". The l/9ᵗʰ Gurkhas were part of the 5ᵗʰ Indian Infantry Brigade of 4ᵗʰ Indian Division engaged in battle close to the town of San Marino in the small principality of the same name.

On the night of 17ᵗʰ/18ᵗʰ September 1944, 1/9 GR was ordered to seize three commanding positions, the last of which was not captured until the early hours of 19ᵗʰ September. By that time the forward company of the battalion had run out of ammunition and it was then that Rifleman Sherbahadur and his section commander stormed an enemy strongpoint killing the machine gunner and putting the remainder of the post to flight. Sherbadadur was awarded the Victoria Cross for his valour in this action.

The official announcement in the London Gazette of 28ᵗʰ December 1944 read as follows:

The KING has been graciously pleased to approve the posthumous award of the VICTORIA CROSS to:
No. 70690 Rifleman Sher Bahadur Thapa, 9ᵗʰ Gurkha Rifles. Indian Army.

"In Italy on the night of 18ᵗʰ/19ᵗʰ September, 1944, a Battalion of the 9ᵗʰ Gurkha Rifles was fighting its way forward into the state of San Marino against bitter opposition from German prepared positions dominating the river valley and held in considerable strength in depth.

Rifleman Sher Bahadur Thapa was a number one Bren gunner in a rifle Company, which just before dawn came under heavy enemy observed small arms and mortar fire. He and his section commander charged an enemy post, killing the machine gunner and putting the rest of the post to flight. Almost immediately another party of Germans attacked the two men and the section commander was badly wounded by a grenade, but, without hesitation, this Rifleman, in spite of intense fire, rushed at the attackers and reaching the crest of the ridge brought his Bren gun into action against the main body of the enemy who were counterattacking our troops.

Disregarding suggestions that he should withdraw to the cover of a slit trench, Rifleman Sher Bahadur Thapa lay in the open under a hail of bullets, firing his Bren gun which he knew he could only bring to bear on the German emplacements from his exposed position on the crest of the hill, as they would not have been visible from the slit trench.

By the intensity and accuracy of the fire which he could bring to bear only from the crest, this isolated Gurkha Bren gunner silenced several enemy machine guns and checked a number of Germans who were trying to infiltrate on to the ridge.

At the end of two hours both forward Companies had exhausted their ammunition and, as they were by then practically surrounded, they were ordered to withdraw. Rifleman Sher Bahadur Thapa covered their withdrawal as they crossed the open ground to positions in the rear and himself remained alone at his post until his ammunition ran out. He then dashed forward under accurate small arms and mortar fire and rescued two wounded men, who were lying between him and the advancing Germans.

While returning the second time he paid the price of his heroism and fell riddled by machine gun bullets fired at point blank range.

The great bravery of this Gurkha soldier was instrumental in saving the lives of many of his companions and his outstanding devotion to duty contributed largely to the severe reverse which the enemy eventually suffered when our troops counter-attacked. His name will live in the history of his Regiment as a very gallant soldier".

On 7th March 1945 at a ceremony outside the Red Fort in Delhi, His Excellency the Viceroy, Field Marshal Lord Wavell, presented her son's Victoria Cross to Sherbahadur's mother, Lachuwa Kumari who had travelled from Nepal especially for the occasion.

RIFLEMAN THAMAN GURUNG VC

5th Royal Gurkha Rifles (Frontier Force)

Date of Action	*Campaign*
10th November 1944	*Italy*

Thaman Gurung was born at Singla, in the district of Gorkha, Western Nepal in 1924. He enlisted in the 5th Royal Gurkha Rifles (Frontier Force) and in September 1944 was posted to the 1st Battalion in Italy then part of 17th Brigade of 8th Indian Division in the Eighth Army.

On 10th November 1944, "A" Company, in which Thaman Gurung was serving, was ordered to send a fighting patrol of one platoon on to Monte San Bartolo, the objective for a future attack. Monte San Bartolo was a high intermediate bluff devoid of cover with steep and precipitous approaches. It was joined to the main feature held by the battalion by a narrow, almost knife-edged, saddle. It was known that this position was occupied by the enemy and the approaches to it covered by a number of machine-gun posts.

It was for his action on this day that Rifleman Thaman Gurung was awarded a posthumous Victoria Cross.

The London Gazette of 22nd February 1945 included the following announcement:

The KING has been graciously pleased to approve the posthumous award of the VICTORIA CROSS to: No. 55537 Rifleman Thaman Gurung, 5th Royal Gurkha Rifles (Frontier Force), Indian Army.

"In Italy on 10th November, 1944, a Company of the 5th Royal Gurkha Rifles was ordered to send a fighting patrol on to Monte San Bartolo, an objective of a future attack. In this patrol were two scouts, one of whom was Rifleman Thaman Gurung.

By skilful stalking both scouts succeeded in reaching the base of the position undetected. Rifleman Thaman Gurung then started to work his way to the summit, but suddenly the second scout attracted his attention to Germans in a slit trench just below the crest, who were preparing to fire with a machine gun at the leading section. Realising that if the enemy succeeded in opening fire, the section would certainly sustain heavy casualties, Rifleman Thaman Gurung leapt to his feet and charged them. Completely taken by surprise, the Germans surrendered without opening fire.

Rifleman Thaman Gurung then crept forward to the summit of the position, from which he saw a party of Germans, well dug in on reverse slopes, preparing to throw grenades over the crest at the leading section. Although the sky-line was devoid of cover and under accurate machine gun fire at close range, Rifleman Thaman Gurung immediately crossed it, firing on the German position with his Tommy gun, thus allowing the forward section to reach the summit, but due to heavy fire from the enemy machine guns, the platoon was ordered to withdraw.

Rifleman Thaman Gurung then again crossed the sky-line alone and, although in full view of the enemy and constantly exposed to heavy fire at short range, he methodically put burst after burst of Tommy gun fire into the German slit trenches, until his ammunition ran out. He then threw two grenades he had with him and rejoining his section, collected two more grenades and again doubled over the bullet-swept crest of the hillock and hurled them at the remaining Germans. This diversion enabled both rear sections to withdraw without further loss.

Meanwhile, the leading section, which had remained behind to assist the withdrawal of the remainder of the platoon, was still on the summit, so Rifleman Thaman Gurung, shouting to the section to withdraw, seized a Bren gun and a number of magazines. He then, yet again, ran to the top of the hill and, although he well knew that his action meant almost certain death, stood up on the bullet-swept summit, in full view of the enemy, and opened fire at the nearest enemy positions. It was not until he had emptied two complete magazines, and the remaining section was well on its way to safety, that Rifleman Thaman Gurung was killed.

It was undoubtedly due to Rifleman Thaman Gurung's superb gallantry and sacrifice of his life that his platoon was able to withdraw from an extremely difficult position without many more casualties than were actually incurred, and very valuable information brought back by the platoon resulted in the whole Monte San Bartolo feature being captured three days later".

Rifleman Thaman Gurung was just 20 years of age when he was killed. On 19th December 1945 at a parade in Delhi, His Excellency the Viceroy, Field Marshal Lord Wavell, presented Thaman Gurung's mother with the Victoria Cross which had been awarded to her son.

His mother, Sardi Gurungseni, was accompanied by her husband, Thaman's stepfather and other members of her family, and went through the trying ceremony with a calmness and dignity in every way worthy of the great and memorable occasion. They had all come from their distant home at Barpak, Gorkha in Western Nepal.

RIFLEMAN BHANBHAGTA GURUNG VC

2nd King Edward VII's Own Gurkha Rifles (The Sirmoor Rifles)

Date of Action
5th March 1945

Campaign
Burma

Rifleman Bhanbhagta was a Gurung, one of the two main tribes from which the 2nd Gurkhas enlisted their men. He was born in September 1921 at Phalpu in the district of Gorkha in Western Nepal and joined the 3rd Battalion, 2nd King Edward VII's Own Gurkha Rifles in 1940, taking part in the first Chindit Expedition in 1943. In 1944 Bhanbhagta was reduced in rank from Naik to Rifleman for something that was subsequently found to be not his fault. It was said that his loss of rank played a part in his determination to prove that he was unjustly treated; in fact he was considered to be a reasonable man and not one to bear a grudge.

The 3/2nd Gurkhas were in 25th Indian Division which in 1944 and 1945 had advanced down the Arakan coast in a number of assault landings. In February 1945 the Battalion landed at Ruywa and, after heavy fighting, advanced to the village of Tamandu. The road from Tamandu to An was dominated by a hill feature known as Snowdon and it was along that road that a Brigade of 82nd West African Division was attempting to evacuate its casualties, 3/2nd Gurkhas were ordered to capture Snowdon in order to let the West Africans evacuate their wounded and this was achieved without opposition. The Japanese however decided to retake an adjoining hill known as Snowdon East and this they did on the night of 4/5th March. "B" Company of 3/2nd Gurkhas was then ordered to recapture lost ground and in their attempt to carry out this task they were subjected to intense enemy fire from their objective and other nearby enemy positions. It was at this stage that one of Bhanbhagta's fellow riflemen was killed in an attempt to rush the enemy positions single-handed and this action of his so inspired the remainder of his Company that, after some three hours, the Japanese were driven from their positions. "B" Company had by then suffered some fifty percent casualties but the enemy casualties had also been substantial.

It was this action that brought a Victoria Cross to Bhanbhagta Gurung. A Gurkha officer of "B" Company expressed the opinion that it was largely the example of Bhanbhagta that inspired the rest of his Company to join him in the final attack which cleared the Japanese off the Snowdon feature, thereby opening the road for the West Africans to evacuate their wounded.

The citation in the London Gazette of 5th June 1945 read:

"In Burma, on 5th March, 1945, a Company of the 2nd Gurkha Rifles attacked an enemy position known as Snowden East. On approaching the objective one of the sections was forced to ground by very heavy Light Machine Gun, grenade and mortar fire, and owing to the severity of this fire was unable to move in any direction. While thus pinned, the section came under accurate fire from a tree sniper some 75 yards to the South. As this sniper was inflicting casualties on the section, Rifleman Bhanbhagta Gurung, being unable to fire from the lying position, stood up fully exposed to the heavy fire and calmly killed the enemy sniper with his rifle, thus saving his section from suffering further casualties.

The section then advanced again, but when within 20 yards of the objective was again attacked by very heavy fire. Rifleman Bhanbhagta Gurung, without waiting for any orders, dashed forward alone and attacked the first enemy foxhole. Throwing two grenades, he killed the two occupants and without any hesitation rushed on to the next enemy fox-hole and killed the Japanese in it with his bayonet.

Two further enemy fox-holes were still bringing fire to bear on the section and again Rifleman Bhanbhagta Gurung dashed forward alone and cleared these with bayonet and grenade. During his single-handed attacks on these four enemy foxholes, Rifleman Bhanbhagta Gurung was subjected to almost continuous and point-blank Light Machine Gun fire from a bunker on the North tip of the objective. Realising that this Light Machine Gun would hold up not only his own platoon which was now behind him, but also another platoon which was advancing from the West, Rifleman Bhanbhagta Gurung for the fifth time went forward alone in the face of heavy enemy fire to knock out this position. He doubled forward and leapt on to the roof of the bunker from where, his hand grenades being finished, he flung two No 77 smoke grenades into the bunker slit. Two Japanese rushed out of the bunker partially blinded by the smoke. Rifleman Bhanbhagta Gurung promptly killed them both with his Khukri. A remaining Japanese inside the bunker was still firing the Light Machine Gun and holding up the advance of No 4 Platoon, so Rifleman Bhanbhagta Gurung crawled inside the bunker, killed this Japanese gunner, and captured the Light Machine Gun.

Most of the objective had now been cleared by the men behind and the enemy driven off were collecting for a counter-attack beneath the North end of the objective. Rifleman Bhanbhagta Gurung ordered the nearest Bren gunner and two Riflemen to take up positions in the captured bunker. The enemy counter-attack followed soon after, but under Rifleman Bhanbhagta Gurung's command the small party inside the bunker repelled it with heavy loss to the enemy.

Rifleman Bhanbhagta Gurung showed outstanding bravery and a complete disregard for his own safety. His courageous clearing of five enemy positions single-handed was in itself decisive in capturing the objective and his inspiring example to the rest of the Company contributed to the speedy consolidation of this success".

His Regiment gained the Battle Honour "Tamandu" and some months later Rifleman Bhanbhagta was decorated with the Victoria Cross at Buckingham Palace by His Majesty King George VI.

After the war his company commander tried to persuade Bhanbhagta to continue serving. However, as he had a frail widowed mother, and a young wife at home in Nepal, he decided to take his discharge. He left his battalion in January 1946, and was given the honorary title of Havildar. Over the years he returned to visit his Regiment in Malaya, Hong Kong and the United Kingdom as an honoured guest.

His three sons followed in his footsteps to serve in the 2nd Battalion of his Regiment but they are now on pension. His own country honoured him in 1945, when he was awarded the Medal of the Order of The Star of Nepal.

His company commander, (later Colonel D F Neill OBE MC) described Bhanbhagta as "a smiling, hard swearing, gallant and indomitable peasant soldier, who, in a battalion of very brave men, was one of the bravest".

Presented to Her Majesty The Queen during her state visit to Nepal in February 1986, he attended many Victoria Cross and George Cross Association reunions in the United Kingdom. On 19th August 1995 he took part in the 'Last Big Parade' in London, the major National event to commemorate the 50th Anniversary of the end of World War 2. He died in Nepal on 1st March 2008.

RIFLEMAN LACHHIMAN GURUNG VC

8th Gurkha Rifles

Date of Action
12th/13th May 1945

Campaign
Burma

Lachhiman Gurung was born at Dakhani in the Chitawan district of Western Nepal. As with most Gurkhas, his actual date of birth is not known but his probable year of birth is 1917. He enlisted into the 8th Gurkha Rifles on 30th December 1940, and was fortunate to be recruited because he was just under five feet in height and, in peacetime this would have meant rejection. Having passed his recruit training, Lachhiman was posted in March 1945 to the 4th Battalion, 8th Gurkha Rifles in Burma.

In May 1945 the 4th Battalion was part of the 89th Indian Infantry Brigade in 7th Indian Division, advancing south along the River Irrawaddy against the retreating Japanese forces. On 11th May, "B" and "C" Companies were ordered to hold an important position astride a track on the west side of the river near to the village of Taungdaw. This track was vital to the Japanese withdrawal, and for three days and nights the Japanese fought with fanatical fury in an attempt to clear the track, launching wave after wave of suicidal attacks.

The key position was held by No 9 Platoon of "C" Company, almost 100 yards ahead of the remainder of the Company, and the most forward post of the platoon was manned by Rifleman Lachhiman's section. This was the situation when Rifleman Lachhiman earned his Victoria Cross. The citation in the London Gazette of 27th July 1945 read:

"At Taungdaw, in Burma, on the west bank of the Irrawaddy, on the night of 12/13th May, 1945, Rifleman Lachhiman Gurung was manning the most forward post of his platoon. At 01.20 hours at least 200 enemy assaulted his Company position. The brunt of the attack was borne by Rifleman Lachhiman Gurung's section and by his own post in particular. This post dominated a jungle path leading up into his platoon locality.

Before assaulting, the enemy hurled innumerable grenades at the position from close range. One grenade fell on the lip of Rifleman Lachhiman Gurung's trench, he at once grasped it and hurled it back at the enemy. Almost immediately another grenade fell directly inside the trench. Again this rifleman snatched it up and threw it back. A third grenade then fell just in front of the trench. He attempted to throw it back, but it exploded in his hand, blowing off his fingers, shattering his right arm and severely wounding him in the face, body and right leg. His two comrades were also badly wounded and lay helpless in the bottom of the trench.

The enemy, screaming and shouting, now formed up shoulder to shoulder and attempted to rush the position by sheer weight of numbers. Rifleman Lachhiman Gurung, regardless of his wounds, fired and loaded his rifle with his left hand, maintaining a continuous and steady rate of fire. Wave after wave of fanatical attacks were thrown in by the enemy and all were repulsed with heavy casualties.

For four hours after being severely wounded Rifleman Lachhiman Gurung remained alone at his post, waiting with perfect calm for each attack, which he met with fire at point-blank range from his rifle, determined not to give one inch of ground.

Of the 87 enemy dead counted in the immediate vicinity of the Company locality, 31 lay in front of this Rifleman's section, the key to the whole position. Had the enemy succeeded in over-running and occupying Rifleman Lachhiman Gurung's trench, the whole of the reverse slope position would have been completely dominated and turned.

This Rifleman, by his magnificent example, so inspired his comrades to resist the enemy to the last, that, although surrounded and cut off for three days and two nights, they held and smashed every attack.

His outstanding gallantry and extreme devotion to duty, in the face of almost overwhelming odds, were the main factors in the defeat of the enemy".

Rifleman Lachhiman Gurung was evacuated to hospital but he lost his right hand and the use of his right eye. On 19th December 1945 he was decorated with the Victoria Cross by His Excellency the Viceroy of India, Field Marshal Lord Wavell, at the Red Fort in Delhi. His proud father, then aged about 74 and very frail, was carried for eleven days from his village in Nepal to be present at the Red Fort to see him decorated.

Lachhiman continued to serve with his regiment, redesignated the 8th Gorkha Rifles after Indian Independence in 1947. He was promoted to Havildar but retired on completion of his service, returning to his village in Nepal. One of his sons subsequently became an Officer in his regiment.

Though severely disabled as a consequence of his war wounds, he travelled to the United Kingdom to be present on the 'Last Big Parade' held in London on 19th August 1995 to commemorate the end of World War 2. He died in the United Kingdom on 12th December 2010 and is buried at Chiswick New Cemetery.

LANCE CORPORAL RAMBAHADUR LIMBU VC

10th Princess Mary's Own Gurkha Rifles

Date of Action	*Campaign*
21st November 1965	*Borneo*

Rambahadur Limbu was born at Chyangthapu in Yangrop Thum, Eastern Nepal in 1939. He is a Limbu which is one of the two principal tribes of eastern Nepal from which men were enlisted by the 7th and 10th Gurkhas of Britain's Brigade of Gurkhas. He was enlisted in November 1957 and joined the 2nd Battalion, 10th PMO Gurkha Rifles, based in Malaya. At that time his battalion was coming to the end of ten years of anti-terrorist operations known as the "Malayan Emergency". Although at the time peace seemed just around the corner, for the Brigade of Gurkhas and many others in the British Army, military operations were never far away.

In December 1962 there was a rebellion in Brunei that was evidence of serious problems in that area, and from 1963 onwards Indonesia brought about "Confrontation" by attempting to make territorial gains in Borneo and Sarawak. During this period 2/10 GR had its full share of the fighting which was mostly at company level.

In November 1965 2/10 GR was ordered to dominate a position some five thousand yards inside the accepted border between Malaysia and Indonesia and it was during this border clash that Lance Corporal Rambahadur Limbu earned the Victoria Cross.

The announcement of the award was contained in the London Gazette of Friday 22nd April 1966 in the following terms:-

The QUEEN has been graciously pleased to approve the award of the VICTORIA CROSS to:
21148786 Lance Corporal RAMBAHADUR LIMBU, 10th Princess Mary's Own Gurkha Rifles.

"On 21st November 1965 in the Bau District of Sarawak Lance Corporal RAMBAHADUR LIMBU was with his Company when they discovered and attacked a strong enemy force located in the Border area. The enemy were strongly entrenched in Platoon strength, on top of a sheer sided hill the only approach to which was along a knife edge ridge allowing only three men to move abreast. Leading his support group in the van of the attack he could see the nearest trench and in it a sentry manning a machine gun. Determined to gain first blood he inched himself forward until, still ten yards from his enemy, he was seen and the sentry opened fire, immediately wounding a man to his right. Rushing forward he reached the enemy trench in seconds and killed the sentry, thereby gaining for the attacking force a first but firm foothold on the objective. The enemy were now fully alerted and, from their positions in depth, brought down heavy automatic fire on the attacking force, concentrating this onto the area of the trench held alone by Lance Corporal RAMBAHADUR LIMBU.

Appreciating that he could not carry out his task of supporting his platoon from this position he courageously left the comparative safety of his trench and, with a complete disregard for the hail of fire being directed at him, he got together and led his fire group to a better fire position some yards ahead. He now attempted to indicate his intentions to his Platoon Commander by shouting and hand signals but failing to do so in the deafening noise of exploding grenades and continuous automatic fire he again moved out into the open and reported personally, despite the extreme dangers of being hit by the fire not only from the enemy but by his own comrades.

It was at the moment of reporting that he saw both men of his own group seriously wounded. Knowing that their only hope of survival was immediate first aid and that evacuation from their very exposed position so close to the enemy was vital he immediately commenced the first of his three supremely gallant attempts to rescue his comrades. Using what little ground cover he could find he crawled forward, in full view of at least two enemy machine gun posts who concentrated their fire on him and which, at this stage of the battle, could not be effectively subdued by the rest of his platoon. For three full minutes he continued to move forward but when almost able to touch the nearest casualty he was driven back by the accurate and intense weight of fire covering his line of approach. After a pause he again started to crawl forward but he soon realised that only speed would give him the cover which the ground could not.

Rushing forward he hurled himself on the ground beside one of the wounded and calling for support from two light machine guns which had now come up to his right in support he picked up the man and carried him to safety out of the line of fire. Without hesitation he immediately returned to the top of the hill determined to complete his self imposed task of saving those for whom he felt personally responsible. It was now clear from the increased weight of fire being concentrated on the approaches to and in the immediate vicinity of the remaining casualty the enemy were doing all they could to prevent any further attempts at rescue. However, despite this Lance Corporal RAMBAHADUR again moved out into the open for his final effort. In a series of short forward rushes and once being pinned down for some minutes by the intense and accurate automatic fire which could be seen striking the ground all round him he eventually reached the wounded man. Picking him up and unable now to seek cover he carried him back as fast as he could through the hail of enemy bullets. It had taken twenty minutes to complete this gallant action and the events leading up to it. For all but a few seconds this young Non-Commissioned Officer had been moving alone in full view of the enemy and under the continuous aimed fire of their automatic weapons. That he was able to achieve what he did against such overwhelming odds without being hit is miraculous. His outstanding personal bravery, selfless conduct, complete contempt of the enemy and determination to save the lives of the men of his fire group set an incomparable example and inspired all who saw him.

Finally rejoining his section on the left flank of the attack Lance Corporal RAMBAHADUR was able to recover the light machine gun abandoned by the wounded and with it won his revenge, initially giving support during the later stages of the prolonged assault and finally being responsible for killing four more enemy as they attempted to escape across the border. This hour long battle which had throughout been fought at point blank range and with the utmost ferocity by both sides was finally won. At least twenty four enemy are known to have died at a cost to the attacking force of three killed and two wounded. In scale and in achievement this engagement stands out as one of the first importance and there is no doubt that, but for the inspired conduct and example set by Lance Corporal RAMBAHADUR at the most vital stage of the battle, much less would have been achieved and greater casualties caused.

He displayed heroism, self sacrifice and a devotion to duty and to his men of the very highest order. His actions on this day reached a zenith of determined, premeditated valour which must count amongst the most notable on record and is deserving of the greatest admiration and the highest praise".

In 1966 Lance Corporal Rambahadur Limbu was presented with his Victoria Cross by Her Majesty the Queen at an Investiture at Buckingham Palace. He was accompanied only by his young son Bhakta, five years old, as his wife had died in the British Military Hospital in Singapore earlier that year.

Rambahadur Limbu became a QGO and, in the rank of Captain (QGO), was appointed to be one of Her Majesty's Gurkha Orderly Officers in 1983. On completion of his duty, he was made MVO and retired in 1985 with the honorary rank of Captain (GCO).

Regularly attending the Victoria Cross and George Cross Association reunions in the United Kingdom, Rambahadur was presented to Her Majesty The Queen in Kathmandu during her state visit to Nepal in February 1986.

On 14th May 2003, Rambahadur took part in the Service at Westminster Abbey when Her Majesty The Queen dedicated the national memorial to holders of the Victoria Cross and George Cross.

The George Cross, Indian Order of Merit, Albert Medal and Empire Gallantry Medal Awards

The George Cross, Indian Order of Merit, Albert Medal and Empire Gallantry Medal.

The narratives that follow cover the awards for exceptional bravery of one George Cross to a Gurkha Non-commissioned Officer, the Indian Order of Merit four times to a Gurkha Officer, five Albert Medals to three British Officers and two Gurkha Riflemen, and the Empire Gallantry Medal to a Gurkha Non-commissioned Officer.

The George Cross (GC) was instituted in 1940 to be awarded "for acts of the greatest heroism or of the most conspicuous courage in circumstances of extreme danger". Intended for award primarily to civilians, awards to military servicemen are confined to actions for which purely military Honours are not normally granted; it is rightly regarded as the 'Civilian VC'. The GC may be awarded posthumously.

The GC subsumed the award of the Medal of the Order of the British Empire (for Gallantry), styled The Empire Gallantry Medal (EGM) and later also subsumed the award of the Albert Medal.

It is appropriate here to outline the historical significance and evolution of the Indian Order of Merit.

In 1837 the Court of Directors of the Honourable East India Company (H.E.I.C) instituted an award for gallantry, the Order of Merit, which pre-dates any British Army gallantry award. In 1860, after the Indian Mutiny, the armed forces of the H.E.I.C were transferred to the Crown and the Order of Merit thereafter became an official award of the British Government. In 1903 the award was designated the 'Indian Order of Merit' (IOM).

A distinctive award and unusual in that, for the period 1837-1911, it had three classes; the 3rd Class awarded for "any conspicuous individual act of gallantry on the part of any native officer or soldier in the field, or in the attack or defence of fortified places, without distinction of rank or grade". For subsequent similar acts of gallantry, the recipient was promoted from the 3rd Class (silver star and laurel wreath) to the 2nd Class (silver star and gold laurel wreath), and for a third award, to the 1st Class (gold star and laurel wreath).

In 1911, when the award of the Victoria Cross was extended "to the Native Officers, Non-commissioned Officers and Men of the Indian Army", the original 1st Class IOM was abolished, the 2nd and 3rd Classes were re-designated 1st and 2nd Class and in 1944 combined into a single Class.

The following narratives include the unique record of a Gurkha Officer who won the IOM for a fourth time when there was no higher honour open to him. Subedar Kishanbir Nagarkoti was awarded the 3rd Class for his first act of gallantry, promoted to 2nd Class and later to 1st Class for further acts of gallantry. His fourth act of gallantry was recognised by the issue of a dated gold bar to be worn on the ribbon; during the 110 years the IOM was awarded, no other soldier achieved that recognition.

The Albert Medal (AM) was instituted in 1866, (being named after the Prince Consort who died in 1861) initially for acts of gallantry in saving life at sea. In 1867, two classes of the medal, 1st and 2nd, were created and ten years later, the awards were extended to "heroic acts performed on land – in preventing loss of life from accidents in mines, on railways, and at fires and other perils on shore".

In 1917, the title of the awards altered; the 1st Class became 'The Albert Medal in Gold' and the 2nd Class 'The Albert Medal'.

In 1949, 'The Albert Medal in Gold' was abolished and replaced by the George Cross and henceforward "The Albert Medal" was only awarded posthumously.

With effect from 31st October 1971, the Royal Warrant instituting the Albert Medal was revoked and holders were invited to exchange their medals for the GC.

The Most Excellent Order of the British Empire instituted in 1917 included an associated medal, eventually styled 'The British Empire Medal' (BEM). In 1922 the BEM was superseded by two new medals, 'The Medal of the British Empire Order (for Gallantry)' and 'The Medal of the British Empire Order (for Meritorious Service)', the medals were awarded in separate Military and Civil Divisions. The former medal was commonly known as, and from 1933 formally designated, 'The Empire Gallantry Medal' (EGM). The living holders of the EGM in 1940 were required to exchange their EGM for the new GC.

The Medal of the British Empire Order (for Meritorious Service) reverted to the style 'The British Empire Medal' (BEM) and was also awarded for gallantry until the Queen's Gallantry Medal (QGM) was instituted in 1974. The BEM continued to be awarded for meritorious service in both Divisions until 1993, when the distinction between the BEM and 5th Class of the Order, the MBE was deemed to be "tenuous and could no longer be sustained"; recommendations for BEMs were discontinued in the United Kingdom and the number of recommendations for MBEs increased.

Some Commonwealth countries continue to include the BEM in their own lists of awards.

NAIK NANDLAL THAPA GC

8th Gurkha Rifles

Date of Action
30th/31st May 1935

Place
Quetta

Nandlal Thapa was born in 1903 in Western Nepal. In 1920 he enlisted in the 8th Gurkha Rifles and was posted to the 2nd Battalion, then at Lansdowne, a hill station in Northern India. Soon after he arrived in the battalion it was ordered to join an expedition to quell a rebellion by the Moplahs, a fanatical group in Southern India. In 1935 the battalion was stationed in Quetta, Baluchistan, and by now Nandlal had been promoted to Naik in command of a rifle section.

In April 1935 the 2/8th Gurkha Rifles had just celebrated a century of their existence, when the following month Quetta experienced the worst earthquake ever known in India. The greatest devastation was in the city and in the civilian residential area; the cantonment area in which both battalions of the 8th Gurkha Rifles were stationed came off comparatively lightly. The initial earthquake occurred on the night of 30th/31st May and as soon as daylight broke all available troops were deployed throughout the area on rescue and salvage work.

Naik Nandlal Thapa's section was one of the first to reach the city area. They soon spotted a number of bodies under the collapsed roof of a building. Tremors were still occurring, debris was falling, and any walls standing were in danger of collapse at any moment. Quite undaunted, and realising the need for immediate action, Naik Nandlal, without a thought for his own safety, dashed into the collapsing building and succeeded in pulling out one of the injured and carrying him to safety. He rushed in again to rescue another injured person, and yet again, while the walls were crumbling about him, to save two more lives. Naik Nandlal and his section worked all day and on their way back to camp he heard a voice crying for help. They stopped and saw a man almost buried under debris. Nandlal and his men dug him out with bare hands. Had it not been for this immediate action, the man would have died within minutes.

For these acts of bravery, Naik Nandlal Thapa was awarded the Empire Gallantry Medal.

The citation in the London Gazette dated 19th November 1935 read as follows:

7506 Naik Nandlal Thapa, 2nd Battalion, 8th Gurkha Rifles, Indian Army.

"Naik Nandlal Thapa formed part of the leading detachment of the Battalion which was conveyed by Mechanical Transport from the Lines to Quetta City. There was no time for the party to collect tools. On arrival in the City the detachment was split up into small parties and worked with their hands for three hours, prior to the distribution of tools, extricating injured men and women from the debris. The area in which they worked was the junction of Bruce Road and Colvin Road, one of the parts most damaged by the earthquake. During this period minor shocks frequently occurred, causing further falls of masonry in the houses in which they were digging with their hands. Working with the party this Non-commissioned Officer showed conspicuous bravery in the manner in which he entered tottering buildings in search of living people and in the work and initiative he displayed in removing them. As a result of his disregard of danger ten people were rescued alive at considerable risk. He fully realised the risks he ran, but was always ready to enter any building where there was any possibility of anybody remaining alive. His conduct was a very fine example of courage and energy at a critical time."

In 1940 King George VI instituted a new decoration, the George Cross (GC), intended as the highest award for gallantry when not under enemy fire. The GC subsumed the Empire Gallantry Medal and living holders were required to exchange their award for a George Cross. Naik Nandlal received a GC in place of his Empire Gallantry Medal.

Nandlal Thapa retired shortly after the Quetta earthquake in the rank of Havildar and returned to his home in Nepal, where he died on 27th June 1987, aged 84.

SUBEDAR KISHANBIR NAGARKOTI IOM

5th Goorkha Regiment
later 5th Royal Gurkha Rifles (Frontier Force)

Date of Action	*Campaign*
18th June 1888	*Black Mountain*

Kishanbir Nagarkoti enlisted in the 5th Goorkha Regiment (The Hazara Goorkha Battalion) during the 1860s. He is first mentioned in the Regimental History for his gallantry during the 2nd Afghan War of 1878-1880, when his Regiment was accorded its first Battle Honours, and particularly for his exploits in the battle of Monghyr Pass on 13th December 1878 when he was awarded the [Indian] Order of Merit, (3rd Class). His campaign medals bear witness to his presence with the Regiment throughout the war, including as they do, the Afghanistan Medal 1878-1880 with the four clasps Peiwar Kotal, Charasia, Kabul and Kandahar, together with the Kabul to Kandahar Star. It was for conspicuous gallantry at the Battle of Charasia on the 6th October 1879 that he gained a 2nd Class [Indian] Order of Merit. In action again at Kabul on 12th December 1879, and now promoted to Naik, he once more distinguished himself, his courage and bravery recognised by admission to the 1st Class of the [Indian] Order of Merit.

On 18th June 1888, by which time Kishanbir Nagarkoti had risen through the ranks to Subedar, he accompanied a small force of the Oghi detachment of the 5th Goorkha Regiment commanded by Major Battye and comprising 58 rifles and 17 police, on an expedition to the eastern slopes of the Black Mountain in the Agror area of the North West Frontier Province. The passage of the troops was resented by the local Gujar tribesmen who constantly harassed the rear of the column. Major Battye held his fire but after failing to persuade the tribesmen that his mission was peaceful, decided to descend towards the village of Atir. Soon afterwards, he received a message stating that the Havildar in charge of the rear-guard had been wounded.

Accompanied by Subedar Kishanbir, Major Battye retraced his steps in order to rescue the wounded Havildar but the main body which was now out of sight knew nothing of this. A stretcher was found for the wounded man but they had scarcely resumed their retirement when the tribesmen, with vastly superior numbers, pressed home their attack.

To cover the withdrawal of the stretcher party, only Major Battye, Captain Urmston of the 6th Punjab Infantry, who was unarmed, Subedar Kishanbir, a Naik, three riflemen and a bugler were now left. Soon the tribesmen attacked this small party. One leaping from cover dealt Captain Urmston a blow with a hatchet. Another, with a sword, severely wounded Major Battye in the shoulder and left him grappling with his assailant. Kishanbir then ran up and killed the tribesman but, with two wounded officers to protect, their retreat cut off, and their small number reduced by casualties, the plight of Subedar Kishanbir's party was desperate. Here, however, Subedar Kishanbir showed superb courage.

He urged his three remaining men to fight on and, using his pistol, killed several of the enemy. Aided by his soldiers, soon reduced to two, he succeeded for a time in keeping the enemy at bay. The end came when Major Battye was killed, shot through the neck, and almost immediately afterwards Captain Urmston received a fatal wound. Surrounded as they were by tribesmen, Kishanbir and his two remaining men could not recover the bodies but they managed to get away themselves and caught up with the main body in the village of Atir. Kishanbir then led the whole detachment back to the scene of his last stand, recovered the bodies, and commanded the return of the force to the safety of Oghi.

All three survivors of the rear-guard party gained awards for their bravery. Sepoys Indarbir Thapa and Motiram Thapa receiving the [Indian] Order of Merit (3rd Class) but an award for Subedar Kishanbir Nagarkoti was more difficult. He had already been awarded the 3rd Class, promoted to the 2nd Class and then to the 1st Class of the Order and initially it was doubtful if the Government could confer any higher distinction. The problem was however resolved by the Gazette announcement covering his award, which read:

"The Governor-General in council is pleased to sanction, as a special case, the grant to Subedar Kishanbir Nagarkoti, 1st Battalion, 5th Goorkha Regiment, Punjab Frontier Force, of a gold bar, with the words "18th June 1888", inscribed thereon, to be attached to and worn with the ribband of the decoration of the First Class of the Order of Merit, in recognition of his conspicuous gallantry on that date on the Black Mountain, Hazara, on which occasion, he, in company with two sepoys of the Regiment, bravely stood by and defended Major Battye and Captain Urmston from the attacks of a numerous body of the enemy".

In 1892 at the age of 44, Subedar Kishanbir retired from the Army and went home to Nepal. In recognition of his exceptional services he was granted a special pension of Rupees 20 a month for life in addition to his normal pension of Rupees 30 a month.

His achievements are unparalleled in the annals of the Indian Army.

His decoration and medals are proudly displayed, together with his portrait in oils, by his Regiment in India, now the 5th Gorkha Rifles (Frontier Force).

OGHI FORT AND THE BLACK MOUNTAIN IN 1858

LIEUTENANT P MALCOLM AM

4th Goorkha Regiment
later 4th Prince of Wales's Own Gurkha Rifles

Date of Action	*Place*
10th June 1887	*Dalhousie (Punjab)*

Pulteney Malcolm was born on the 16th August 1861, the son of General Sir George Malcolm, GCB. He was educated at Wellington College and the Royal Military College Sandhurst. Commissioned on 11th August 1880 into The Royal Fusiliers, he served in India with the 2nd Battalion in 1881/82 and 1884/85.

In 1886 he was appointed ADC to the Commander-in-Chief, Bombay [Presidency] Army and later that year transferred to the newly raised 2nd Battalion of the 4th Goorkha Regiment, (later 4th Prince of Wales's Own Gurkha Rifles).

On 10th June 1887 he attempted to save the life of an officer in The West Yorkshire Regiment who had fallen over a precipice in Dalhousie; for this action he was awarded the Albert Medal, Second Class. The citation published in the London Gazette of 25th September 1888 read as follows:

"The Queen has been graciously pleased to confer, THE ALBERT MEDAL OF THE SECOND CLASS' upon Lieutenant Pulteney Malcolm, 4th Goorkha Regiment, in recognition of the conspicuous gallantry displayed by him on the 10th June 1887, in attempting to save the life of a comrade who had fallen over a precipice near Dalhousie, East India".

Subsequently he took part in the Chin-Lushai Expedition of 1889-90 and was promoted Captain on 11th August 1891. In 1892 he was appointed to the staff of the Meerut Division and in 1895 rejoined his battalion for operations with the Chitral Relief Force.

In 1897 he served on the staff of the Force despatched to the Tochi Valley in Waziristan and then rejoined his battalion for active service with the Tirah Field Force and was Mentioned in Despatches. Further staff appointments followed in 1898 and 1899 and in 1901 he was appointed Chief Staff Officer in the Waziristan Field Force. During operations he had his horse shot from under him, was Mentioned in Despatches and in September 1902, awarded the DSO.

In 1903 Major Malcolm returned to England on leave pending retirement and took up police service. On 17th December 1903 he was appointed Head Constable of Kingston-on-Hull and on the 30th September 1910, Chief Constable of Cheshire. During the visit of King George V to Staffordshire and Cheshire in April 1913 he was appointed MVO (4th Class).

During World War 1 he volunteered for active service in the Army, was promoted to Lieutenant Colonel and in 1915 joined the Staff of the 2nd London Division. Serving in France he was again Mentioned in Despatches but had to be invalided home in late 1916 and was discharged. He resumed the appointment of Chief Constable of Cheshire in January 1917.

Awarded the King's Police Medal in 1925 and appointed CBE (Civil Division) in the King's Birthday Honours List 1932, he retired on 30th April 1934 after more than half a century of distinguished military and civil service.

Lieutenant Colonel Pulteney Malcolm CBE DSO MVO AM KPM died in London on 20th April 1940; a funeral service was held at Norwood Crematorium and an internment at Westerkirk Langholm.

no picture available

CAPTAIN D C YOUNG AM

4th Gurkha Rifles
later 4th Prince of Wales's Own Gurkha Rifles

Date of Action	*Place*
30th August 1906	*Ferozepore*

The son of Colonel David Butler Young, David Coley Young was born on 5th December 1869, was commissioned on 21st September 1889, and joined the 1st Battalion 4th Gurkha (Rifle) Regiment (later 4th Prince of Wales's Own Gurkha Rifles) in 1892.

He was promoted Lieutenant in February 1892, and took part in the Waziristan campaign in 1894-95. He was appointed Adjutant in 1898 and promoted Captain on 21st September 1900. He accompanied the Battalion to China in 1900-01.

In 1913, he was awarded the Albert Medal for saving life during a fire at the Ferozepore Arsenal in 1906.

The citation which appeared in the London Gazette of 26 August 1913 read as follows:

"The King has been pleased to approve of Albert Medals being conferred upon the under mentioned officers and non-commissioned officer in recognition of their gallantry in saving life on the occasion of a fire caused by explosions of cordite at Ferozepore in the year 1906.

Albert Medal of the Second Class
Major (then Captain) David Coley Young, Captain (then Lieutenant) Basil Condon Battye R E, and Staff-Sergeant (then Corporal) Patrick John Fitzpatrick."

A full description of the explosion and of the gallantry of various officers and others to whom Albert Medals were awarded in 1913 will be found in the London Gazette of September 26th, 1911. The following is an extract:

"Captain Ross discovered the fire, and with a detachment of his regiment entered the magazine compound with a small hand engine fed from tanks in the magazine, and attempted to put out the fire. He also worked at getting the steam engine into position.

Major Young, as General Anderson's Brigade-Major, was constantly with the General in positions of great danger. In particular he joined General Anderson at a critical moment by the door of No. 8 cell, from which the gunpowder was being removed, and remained with the General throughout the rest of the period of danger.

Captain Battye assisted in the removal of the gunpowder from No. 8 cell. He also, with Staff Serjeant Fitzpatrick, directed the operations for piercing two holes through the masonry of the roof of Cell No. 9, where the small arms ammunition was burning, and succeeded in getting the hose through these holes so as to play on the burning ammunition. By this means a check on the fire in No 9 was effected. Both men were conspicuous throughout the day in the magazine enclosure".

Having been promoted Major on 21st September 1907, David Young served as a Double Company Commander in the 2nd Battalion 3rd Gurkha Rifles in December 1914; in January 1915 he was transferred to the 1st Battalion 4th Gurkha Rifles in

France, again as a Double Company Commander.

It was while he was commanding the Battalion at Neuve Chapelle on 12th March 1915 that he was killed while attempting to recover a wounded man of The Leicestershire Regiment who was lying in front of the Battalion trenches. He was Mentioned in Despatches after the battle for "conspicuous gallantry and devotion to duty".

Lieutenant Colonel D C Young is buried in the Rue-des-Berceaux Military Cemetery at Richebourg-L'Avoue in France.

JAMRUD FORT

CAPTAIN (TEMPORARY MAJOR) C L N NEWALL AM

2nd King Edward's Own Gurkha Rifles (The Sirmoor Rifles)
later 2nd King Edward VII's Own Gurkha Rifles (The Sirmoor Rifles)

Date of Action	*Place*
3rd January 1916	*France*

Cyril Louis Norton Newall was born on 15th February 1886, the son of Lieutenant Colonel W P Newall, 2nd Goorkhas, who commanded the 1st Battalion 2nd (The Prince of Wales's Own) Gurkha (Rifle) Regiment (The Sirmoor Rifles) from 1894 to1897. He was educated at Bedford School and the Royal Military College, Sandhurst

He was commissioned on 16th August 1905 into The Royal Warwickshire Regiment, in which Regiment he was promoted Lieutenant in 1907 and served in the Zakka Khel expedition of 1908. He transferred to his father's Regiment and joined the 2nd Battalion of the 2nd Gurkha Rifles, in 1909. He qualified as a 'Pilot Aviator' on 3rd October 1911, having learnt to fly in a Bristol Biplane at Larkhill whilst he was in England on leave. He was appointed Quartermaster of the 2nd Battalion 2nd Gurkha Rifles in 1911 and Adjutant on 15th March 1913. Having gained his RFC 'wings' at the Central Flying School, Upavon in 1913, he was, when World War 1 broke out, a Captain and an instructor at the Indian Central Flying School at Sitapur.

He returned to the UK firstly to become a Flight Commander in 1 Squadron and then to lead 12 Squadron and take it to France, where, equipped with the BE2c aircraft, it took part in the battle of Loos, bombing railways and carrying out reconnaissance.

In 1916 Newall was serving in France when a bomb store caught fire and for his actions on that day, he was awarded the Albert Medal, First Class. The citation read as follows:

"On 3rd January 1916 a fire broke out inside a large Royal Flying Corps bomb store at St Omer, France. The store held nearly 2,000 high explosive bombs, some with very large charges, and a considerable number of incendiary bombs which were burning freely.

Major Newall at once took all the necessary precautions and then, assisted by Air Mechanic Simms, poured water into the shed through a hole which had been burnt in the wall. He then sent for the key to the shed and with Simms and two other R.F.C personnel entered the building and succeeded in extinguishing the fire at great personal risk to all four men. It was discovered afterwards that the wooden cases containing the bombs had all been burnt, with some completely destroyed, and a major explosion had been imminent".

Thereafter, Newall was marked for distinction and his steady rise was impressive. By the end of 1916 he was Commander 9 Wing, whose seven squadrons provided the RFC's main long range bombing and reconnaissance force in France. In October 1917 he was appointed to lead the new 41 Wing, which had been formed as part of the counter measures to the German night bombing of London. Based at Ochey, near Nancy in Eastern France, the Wing's specific objective was to attack targets of military importance in Germany. In February 1918 the Wing increased in size to become VIII Brigade and by June had carried out 142 raids, 57 of them in Germany. With the decision to expand the Brigade into the Independent Force under Trenchard, Newall now became his deputy but the war was over before its potential could be realised. He was Mentioned in Despatches three times, and awarded the French Legion d'Honneur, the Belgian Croix de Guerre, and the Order of the Crown of Italy 4th

Class, as well as the Albert Medal.

After the war Newall spent three years in the Air Ministry as Deputy Director of Personnel and in 1919, having been appointed a Companion of the Order of St. Michael and St. George (CMG) and a Commander of the Order of the British Empire (CBE), was transferred to the newly formed Royal Air Force and promoted to Group Captain. In 1922 he became deputy to the Commandant of the Technical Training School at Halton. He was made an ADC to the King 1923-24, promoted Air Commodore in 1925, and from 1926 to 1931 served in the Air Ministry as Director of Operations and Intelligence, and Deputy Chief of the Air Staff. In 1929 he was made a Companion of the Order of the Bath (CB) and in 1930 promoted to Air Vice Marshal. From 1931 to 1934 he was Air Officer Commanding RAF Middle East and in 1935 promoted to Air Marshal and advanced to KCB. He was Chief of the Air Staff from 1937 to 1940, was promoted Air Chief Marshal in 1937, and promoted to GCB in 1938. In 1940 he became Marshal of the Royal Air Force and was made a member of the Order of Merit (OM). In 1941 he was promoted to GCMG and appointed a Knight of the Order of St. John of Jerusalem. At the end of 1941 he became Governor General and Commander-in-Chief of New Zealand and served there until 1946 when he returned home and was created 1st Baron Newall of Clifton-upon-Dunsmoor, in the county of Warwick.

Trenchard, Dowding, and Slessor were all contemporaries of Newall and it was the latter who once remarked of Newall that he "had seldom met a man who was so good for one's morale" and who was, in his opinion "the prime architect of the wartime Air Force".

Lord Newall married twice, his first wife dying in 1924, and had one son and a daughter. He died on 30th November 1963.

no picture available

RIFLEMAN AIMANSING PUN AM

6th Gurkha Rifles
later 6th Queen Elizabeth's Own Gurkha Rifles

Date of Action	*Place*
16th May 1926	*Attock (near Peshawar)*

At the end of 1925 the 1st Battalion 6th Gurkha Rifles took part in extensive Northern Command manoeuvres in the Indus valley, near Attock in the north-west frontier of India.

In May of 1926, 3790 Rifleman Aimansing Pun made a most gallant attempt to save a comrade from drowning at the confluence of the Kabul and Indus Rivers for which he was later awarded the Albert Medal.

The citation from the London Gazette of 21st September 1926 read as follows:

"On 16th May 1926, a party of men were washing their clothes on the banks of a wide river which, by reason of the strong converging currents, was extremely dangerous to swimmers. Contrary to orders, one of the party, Kishen Bahadur Thapa, entered the stream, swam out some fifty yards from the bank, was caught by the current and rendered helpless. Although well acquainted with the danger involved, Rifleman Aimansing Pun, Gurkha Rifles, without hesitation, went to the rescue of his comrade and succeeded in getting hold of the drowning man. He commenced to swim with him to the further bank, but his efforts proved ineffectual owing to the violent struggles of Kishen Bahadur Thapa, who dragged his rescuer under water. On coming to the surface Pun, whose hold on the drowning man had been broken, could not see his comrade. He reached land only with great difficulty."

At the time of the incident, Aimansing was serving in 'C' Company; the Company Commander was Major H R K Gibbs whose initial report culminated in the award. In 1947 Lieutenant Colonel Gibbs was serving in the Gurkha Recruiting Depot at Kunraghat and interviewed a Gurkha who turned out to be Aimansing's brother. As a result of this fortuitous meeting, Aimansing's death was established and his Albert Medal gifted to the Regiment.

A bronze award for Saving Life on Land, the Albert Medal is now in the Gurkha Museum Collection. The reverse of the medal is engraved, "Presented in the name of His Majesty to Rifleman Aimansing Pun 1/6th Gurkha Rifles for Gallantry in endeavouring to save life at Attock on 16th May 1926".

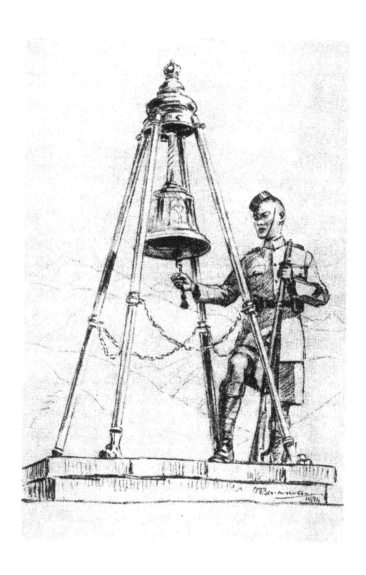

no picture available

RIFLEMAN HARKBIR THAPA AM

8th Gurkha Rifles

Date of Action	*Place*
31st May 1935	*Quetta*

In May 1935, Rifleman Harkbir Thapa was serving with the 2nd Battalion 8th Gurkha Rifles in Quetta, when a severe earthquake shook the town. Rifleman Harkbir Thapa was one of a number of 8th Gurkhas who risked their lives in giving assistance on that day and for which he was awarded the Albert Medal. The citation from the London Gazette of 19th November 1935 read as follows:

"On the morning of 31st May Rifleman Harkbir Thapa, 8th Gurkha Rifles, was detailed as part of a rescue party which was going to dig out some living people behind the Police Lines. On the way to the work he heard noises in a building and obtained permission from the N.C.O. to try and get these people out. At about 6.30 am the Adjutant visited the area to ascertain how work was progressing. He found Rifleman Harkbir Thapa had worked his way, with his hands, through the debris under a tottering roof, and was rescuing two people who were alive but buried. As there was clearly every chance of the roof collapsing onto him as he removed the debris the Adjutant assisted him by propping up the roof as far as possible. Rifleman Harkbir Thapa continued his work and brought out two children alive. He undoubtedly saved these two children at the risk of his own life.

On 2nd June this Rifleman's conduct was again brought to notice. On this occasion he formed part of a detachment working in Hudda Village. The upper storey of a crumbling house was being cleared, part of the roof had fallen through into a lower storey, thus rendering the floor most dangerous. A living child was discovered in the lower storey. This man and one other volunteered to dig through a corner of the floor opposite to where it had crumbled. They did so with khukries and their hands and got through to the lower storey and rescued the child. They did this at considerable risk to their own lives, as the walls were in danger of falling and the floor might have collapsed at any moment."

Harkbirs Record of Service has not been found.

LANCE NAIK CHITRABAHADUR GURUNG EGM

8th Gurkha Rifles

Date of Action
30ᵗʰ/31ˢᵗ May 1935

Place
Quetta

Chitrabahadur Gurung was born in Western Nepal about 1912. He was recruited into the 1ˢᵗ Battalion, 8ᵗʰ Gurkha Rifles and joined it in Shillong, Assam, in 1929. This was the first occasion on which the two battalions of the 8ᵗʰ Gurkha Rifles had served together in their long history; the second time was in Quetta in 1935, where Chitrabahadur was destined to make a name for himself.

During the night of 30ᵗʰ/31ˢᵗ May 1935, a disastrous earthquake occurred in Quetta, the worst one India had known. Chitrabahadur Gurung had by now been promoted to Lance Naik and was commanding a rifle section. Together with all other available men, he was sent out at first light on 31ˢᵗ May on rescue and salvage work, his section being sent to the civilian residential area that, together with the city, had suffered the most damage. His first task was at Railway Police Superintendent Newington's bungalow. This had collapsed and he and his wife were under the debris.

What happened next is shown in the following citation that appeared in the London Gazette on 19ᵗʰ November 1935:

No 985 Lance-Naik Chitrabahadur Gurung, 1ˢᵗ Battalion, 8ᵗʰ Gurkha Rifles, Indian Army.

"This Non-commissioned Officer, along with other men, was instrumental in saving the life of Mrs Newington (wife of Superintendent Newington of the Railway Police). The family of Mr and Mrs Newington were buried in the ruins of their bungalow some five or six feet under the debris. The party started digging at about 07.00 hours on 31ˢᵗ May, 1935. Through unceasing work carried out under very dangerous circumstances, as the work was carried out under the remaining wall of the bungalow, which would have fallen if another shock had taken place, the vicinity of the bodies was reached about 09.30 hours. Great care and intelligence were required in order to avoid wounding the couple, whilst speed was essential if there was to be any hope of rescue. Soon after 09.30 hours Mrs Newington was extricated alive and shortly after Mr Newington dead. Lance-Naik Chitrabahadur Gurung showed initiative, resource, and courage, which were undoubtedly instrumental in saving life."

Lance Naik Chitrabahadur Gurung and his section carried on with rescue work without rest for the remainder of the day.

For his outstanding work, courage and initiative on that day, Lance Naik Chitrabahadur Gurung was awarded the Empire Gallantry Medal. In 1940, King George VI instituted the George Cross as the highest award for acts of conspicuous courage in circumstances of extreme danger; the decoration subsumed the award of the Empire Gallantry Medal and living holders had their medals replaced with the new decoration. Lance Naik Chitrabahadur Gurung had died in 1939 and his medal was never exchanged; his Empire Gallantry Medal is now held in the Gurkha Museum Collection of Orders, Decorations and Medals.

THE GURKHA MUSEUM PUBLICATIONS

Publications issued so far:

Number   Title

1.   The Story of Gurkha VCs.
ISBN 978-1-908487-33-9
November 2012. 96 pages + cover

2.   The Lineages and Composition of Gurkha Regiments in British Service.
March 1978, Revised editions May 1982,. September 1984, May 1997.
Second Edition, March 2010. 214 pages +5 title pages

3.   Campaign and Service Medals Awarded to Gurkha Regiments in British Service.
March 1978. Revised editions July 1985, February 1987, April 1987, September 1991 and
November 1996. 19 pages + Cover

4.   Bibliography of Gurkha Regiments and Related Subjects. April 1980.
Revised editions, April 1982, July 1985 and April 1987. Second edition May 1994.
Third edition January 2006. 136 Pages + Cover

5.   Recipes From the Brigade of Gurkhas. October 1991. Revised edition March 1994.
Reset and reprinted August 1998. 38 Pages + Cover

6.   Insignia of The 1st King George V's Own Gurkha Rifles (The Malaun Regiment).
January 1987, Revised editions August 1988, August 1989, June 1996, May 2000 and
May 2002. 53 Pages + Cover

7.   Insignia of The 8th Gurkha Rifles.
November 1996. Revised editions February 1998, May 2000 and May 2002. 57 Pages + Cover

8.   Insignia of The 6th Queen Elizabeth's Own Gurkha Rifles.
December 1997. Revised editions May 2000 and May 2002. 66 Pages + Cover

9.   Insignia of The 2nd King Edward VII's Own Gurkha Rifles (The Sirmoor Rifles).
July 1998. Revised editions January 1999 and August 2007. 99 Pages + Cover

10.   Insignia of The 7th Duke of Edinburgh's Own Gurkha Rifles.
March 2000. Revised edition May 2002. 90 Pages + Cover

11.   Insignia of The 9th Gurkha Rifles.
August 2000. Revised editions February 2001 and May 2002. 46 Pages + Cover

12.   Insignia of The 10th Princess Mary's Own Gurkha Rifles.
March 2006. 100 pages + Cover

13.   Insignia of The 3rd Queen Alexandra's Own Gurkha Rifles.
August 2002. Revised edition August 2007. 62 Pages + Cover

14.   Insignia of The 4th Prince of Wales's Own Gurkha Rifles.
January 2006. 61 Pages + Cover

15.   Insignia of The 5th Royal Gurkha Rifles (Frontier Force).
March 2006. 56 Pages + Cover

16.   Insignia of Gurkha Units in the British Army raised after 1947 being written.

17.   Insignia of Gurkha Units in India and Burma
(Para-Military, Military Police, Frontier Force and War-raised units) up to 1948.
March 2006. 75 Pages + Cover

18.   The Story of Gurkha Hill Racing.
2009. 5 Pages + Cover

19.   The Scottish Connection and Gurkha Pipe Bands.
2010. 23 Pages + Cover

20.   Battle Honours Awarded to Gurkha Regiments and Related Subjects March 1978.
Revised editions July 1985, February 1987 and November 1996. 42 Pages + Cover

21.   Order of Battle of Gurkha Units 1940 – 1946 Showing the Formations and Theatres of
War in Which They Served.
June 1991. 23 Pages + cover

22.   The Gurkha Brigade in The Great War 1914 – 1920. A Brief Record of The Service of
Each Unit and of The Formations and Theatres of War in Which They Served.
July 1991. 19 Pages + Cover

23.   Gurkhas at Gallipoli May – December 1915.
May 1992. 19 Pages + Cover

24.   Gurkha Battalions in Italy 1943 – 1945.
           32 Pages + Cover
25.   The Gurkhas in World War 2 - 4th Indian Division in North Africa 1942 – 43.
           5 Pages + Cover
26.   The 4th Indian Divisional Signals at Cassino 14 February – 27 March 1944.
           2009. 34 Pages + Cover
27.   The Gurkhas in World War 2 - Chindits.
           2009. 10 pages + Cover
28.   Gurkhas in the Burma Campaign 1941 – 1946. 1994.
           16 Pages + Cover
29.   The Gurkha Parachutist.
           2009. 15 Pages + Cover
30.   The Second Afghan War 1878-80. The Third Afghan War 1919.
           2009. 18 Pages + Cover
31.   The North West Frontier 1919 – 1947 – Two Frontier Actions.
           2009. 14 Pages + Cover
32.   Images of Delhi 1857.
           2007. 44 Pages + Cover
33.   Gurkhas in the Indian Mutiny 1857 – 1859.
           2007. 23 pages + cover
34.   The Royal Flying Corps and Royal Air Force in the North West Frontier 1915 – 1941.
           27 Pages + Cover
35.   North West Frontier, The Pathan Origins.
           10 Pages + Cover
36.   11th Gurkha Rifles 1918 – 1922.
           17 Pages + Cover

Available from

The Gurkha Museum
Peninsula Barracks
Romsey Road
Winchester SO23 8TS
England

9781908487339

ND - #0147 - 270225 - C0 - 297/210/5 - PB - 9781908487339 - Gloss Lamination